Lecture Notes in Geoinformation and Cartography

Series Editors

William Cartwright, Department of Land Information, RMIT University, Melbourne, VIC, Australia

Georg Gartner, Department of Geodesy and Geoinformation, Vienna University of Technology, Vienna, Austria

Liqiu Meng, Lehrstuhl für Kartographie, TU München, München, Bayern, Germany

Michael P. Peterson, Department of Geography and Geology, University of Nebraska at Omaha, Omaha, NE, USA

The Lecture Notes in Geoinformation and Cartography series provides a contemporary view of current research and development in Geoinformation and Cartography, including GIS and Geographic Information Science. Publications with associated electronic media examine areas of development and current technology. Editors from multiple continents, in association with national and international organizations and societies bring together the most comprehensive forum for Geoinformation and Cartography.

The scope of Lecture Notes in Geoinformation and Cartography spans the range of interdisciplinary topics in a variety of research and application fields. The type of material published traditionally includes:

- proceedings that are peer-reviewed and published in association with a conference;
- post-proceedings consisting of thoroughly revised final papers; and
- research monographs that may be based on individual research projects.

The Lecture Notes in Geoinformation and Cartography series also includes various other publications, including:

- tutorials or collections of lectures for advanced courses;
- contemporary surveys that offer an objective summary of a current topic of interest; and
- emerging areas of research directed at a broad community of practitioners.

More information about this series at https://link.springer.com/bookseries/7418

Rodrigo Tapia-McClung · Oscar Sánchez-Siordia ·
Karime González-Zuccolotto ·
Hugo Carlos-Martínez
Editors

Advances in Geospatial Data Science

Selected Papers from the International
Conference on Geospatial Information
Sciences 2021

 Springer

Editors
Rodrigo Tapia-McClung (iD)
Centro de Investigación en Ciencias de
Información Geoespacial - CDMX
Tlalpan, Mexico City, Mexico

Karime González-Zuccolotto
Centro de Investigación en Ciencias de
Información Geoespacial - CDMX
Tlalpan, Mexico City, Mexico

Oscar Sánchez-Siordia
Centro de Investigación en Ciencias de
Información Geoespacial - Yucatán
Parque Científico Tecnológico Yucatán
(PCTY)
Mérida, Yucatán, Mexico

Hugo Carlos-Martínez
Centro de Investigación en Ciencias de
Información Geoespacial - Yucatán
Parque Científico Tecnológico Yucatán
(PCTY)
Mérida, Yucatán, Mexico

ISSN 1863-2246   ISSN 1863-2351 (electronic)
Lecture Notes in Geoinformation and Cartography
ISBN 978-3-030-98098-6   ISBN 978-3-030-98096-2 (eBook)
https://doi.org/10.1007/978-3-030-98096-2

This Springer imprint is published by the registered company Springer Nature Switzerland AG
The registered company address is: Gewerbestrasse 11, 6330 Cham, Switzerland

Organization

iGISc 2021 was organized by the Center of Research in Geospatial Information Sciences, Mexico, and the National Geointelligence Laboratory (GeoInt).

Program Committee Members

Olga Lidia Acosta-López, Pontificia Universidad Católica de Chile
Pedro Camilo Alcántara-Concepción, Universidad de Guanajuato, Mexico
Giner Alor-Hernández, Instituto Tecnológico de Orizaba, Mexico
Hugo Carlos-Martínez, Centro de Investigación en Ciencias de Información Geospacial, Mexico
Mario Chirinos-Colunga, Centro de Investigación en Ciencias de Información Geospacial, Mexico
Gustavo Cruz-Bello, Universidad Autónoma Metropolitana, Mexico
Michelle Farfán-Gutiérrez, Universidad de Guanajuato, Mexico
Yan Gao, Centro de Investigaciones en Geografía Ambiental, Universidad Nacional Autónoma de México, Mexico
Graciela González-Farías, Centro de Investigación en Matemáticas, Mexico
Karime González-Zuccolotto, Centro de Investigación en Ciencias de Información Geospacial, Mexico
Gandhi Hernández-Chan, Centro de Investigación en Ciencias de Información Geospacial, Mexico
Mikel Iruskieta, University of the Basque Country
Lilián Juárez-Téllez, Centro de Investigación en Ciencias de Información Geospacial, Mexico
Ivan López-Arevalo, Centro de Investigación y de Estudios Avanzados del Instituto Politécnico Nacional, Tamaulipas, Mexico Marco Antonio López-Vega, Instituto de Geografía, Universidad Nacional Autónoma de México, Mexico
Jean Francois Mas, Centro de Investigaciones en Geografía Ambiental, Universidad Nacional Autónoma de México, Mexico

María Elena Méndez-López, Centro de Investigación en Ciencias de Información Geospacial, Mexico
Alejandro Molina-Villegas, Centro de Investigación en Ciencias de Información Geospacial, Mexico
Luis Alberto Muñoz-Ubando, Universidad Autónoma de Yucatán, Mexico
Rosa Martha Peralta-Blanco, Centro de Investigación en Ciencias de Información Geospacial, Mexico
Alejandro Rodríguez-González, Universidad Politécnica de Madrid, Spain
José Luis Sánchez-Cervantes, Instituto Tecnolóóico de Orizaba, Mexico
Oscar S. Siordia, Centro de Investigación en Ciencias de Información Geospacial, Mexico
José Luis Silván-Cárdenas, Centro de Investigación en Ciencias de Información Geospacial, Mexico
Rodrigo Tapia-McClung, Centro de Investigación en Ciencias de Información Geospacial, Mexico
CarlosVilalta, Centro de Investigación en Ciencias de Información Geospacial, Mexico

Workshop and Session Organizing Chairs

Karime González-Zuccolotto, Centro de Investigación en Ciencias de Información Geospacial, Mexico
Lilián Juárez-Téllez, Centro de Investigación en Ciencias de Información Geospacial, Mexico
María Elena Méndez-López, Centro de Investigación en Ciencias de Información Geospacial, Mexico
Rodrigo Tapia-McClung, Centro de Investigación en Ciencias de Información Geospacial, Mexico

Publicity Committee Chairs

Karime González-Zuccolotto, Centro de Investigación en Ciencias de Información Geospacial, Mexico
Lilián Juárez-Téllez, Centro de Investigación en Ciencias de Información Geospacial, Mexico
María Elena Méndez-López, Centro de Investigación en Ciencias de Información Geospacial, Mexico
Rosa Martha Peralta-Blanco, Centro de Investigación en Ciencias de Información Geospacial, Mexico
Marisol Sosa-Padilla, Centro de Investigación en Ciencias de Información Geospacial, Mexico

Reviewers

Olga Lidia Acosta-López, Pontificia Universidad Católica de Chile
Pedro Camilo Alcántara-Concepción, Universidad de Guanajuato, Mexico
Giner Alor-Hernández, Instituto Tecnológico de Orizaba, Mexico
Carlos Brito-Loeza, Centro de Incvestigación Científica de Yucatán, Mexico
Hugo Carlos-Martínez, Centro de Investigación en Ciencias de Información Geospacial, Mexico
Camilo Alberto Caudillo-Cós, Centro de Investigación en Ciencias de Información Geospacial, Mexico
Mario Chirinos-Colunga, Centro de Investigación en Ciencias de Información Geospacial, Mexico Robert Constantinescou, University of South Florida, United States of America
Gustavo Cruz-Bello, Universidad Autónoma Metropolitana, Mexico
Arturo Espinosa-Romero, Universidad Autónoma de Yucat'an, Mexico
Michelle Farfán-Gutiérrez, Universidad de Guanajuato, Mexico
Dolors Ferrés, Escuela Nacional de Ciencias de la Tierra, Universidad Nacional Autónoma de México, Mexico
Yan Gao, Centro de Investigaciones en Geografía Ambiental, Universidad Nacional Autónoma de México, Mexico
Graciela González-Farías, Centro de Investigación en Matemáticas, Mexico
Karime González-Zuccolotto, Centro de Investigación en Ciencias de Información Geospacial, Mexico
Gandhi Hernández-Chan, Centro de Investigación en Ciencias de Información Geospacial, Mexico
Mikel Iruskieta, University of the Basque Country
Mariana Patricia Jácome-Paz, Instituto de Geofísica, Universidad Nacional Autónoma de México, Mexico
Lilián Juárez-Téllez, Centro de Investigación en Ciencias de Información Geospacial, Mexico
Ivan López-Arevalo, Centro de Investigación y de Estudios Avanzados del Instituto Politécnico Nacional, Tamaulipas, Mexico Marco Antonio López-Vega, Instituto de Geografía, Universidad Nacional Autónoma de México, Mexico
Pablo López-Ramírez, Centro de Investigación en Ciencias de Información Geospacial, Mexico
Jean Francois Mas, Centro de Investigaciones en Geografía Ambiental, Universidad Nacional Autónoma de México, Mexico
María Elena Méndez-López, Centro de Investigación en Ciencias de Información Geospacial, Mexico
Jorge Alberto Montejano-Escamilla, Centro de Investigación en Ciencias de Información Geospacial, Mexico
Alejandro Molina-Villegas, Centro de Investigación en Ciencias de Información Geospacial, Mexico
Luis Alberto Muñoz-Ubando, Universidad Autónoma de Yucatán, Mexico

Preface

This volume consists of the selected peer-reviewed papers from the International Conference on Geospatial Information Sciences 2021 that took place on November 3–5, 2021. Initially planned to be held in person in Mérida, Yucatán, Mexico, due to the coronavirus pandemic it was decided to be an online event. These papers were selected by the Scientific Program Committee of the Conference after a peer-review process. They represent the vast scope of the interdisciplinary research areas that characterize the Geospatial Information Sciences. It represents a fabulous opportunity to showcase research carried out by young researchers, especially Mexican ones, show it to the rest of the world, and enhance the growth of the Sciences in the country.

iGISc 2021 is the second iteration of a successful conference that aims at bringing together international experts in the field of Geospatial Information Sciences (GISc) to foster the exchange of ideas, knowledge, and experiences. The conference hosts a broad array of subjects related to the acquisition, processing, modeling, analysis, visualization, and use of Geographic Information.

As an emergent conference in the country, the first edition brought together little less than 100 experts and students in the field. In contrast, the online version of the conference attracted more than 300 participants. Apart from the oral presentations of the selected papers, the conference boasted an interesting offer of workshops covering different topics and aspects of the use of Geospatial Information Sciences. Additionally, several keynote speakers were invited to share their knowledge and insights with the participants.

Please visit the iGISc webiste at http://igisc.org where you can find all about this interesting event.

Mexico City, Mexico
Mérida, Yucatán, Mexico
Mexico City, Mexico
Mérida, Yucatán, Mexico
November 2021

Rodrigo Tapia-McClung
Oscar Sánchez-Siordia
Karime González-Zuccolotto
Hugo Carlos-Martínez

Acknowledgements

iGISc 2021 was a successful conference thanks to the authors, presenters, participants, keynote speakers, workshop organizers, session chairs, organizing committee members, student volunteers, and reviewers. Thank you all for your help, commitment, and support in making this a successful event.

Special thanks go to our keynote speakers, Mateo Valero (Barcelona Supercomputing Center, Universitat Politècnica de Catalunya), Alberto Giordano (Texas State University), Ulises Cortés (Barcelona Supercomputing Center, Universitat Politècnica de Catalunya), Alison Heppenstall (University of Leeds), Andrew Crooks (University at Buffalo), and Daniel Arribas-Bel (University of Liverpool), as well as the workshop presenters.

We would also like to thank all the presenters and participants who put a lot of effort in being part of our *Gather* community. While trying to adjust to time differences all over the globe, we are grateful for those participants who were able to catch the presentations live, as this year's participants came from countries across different time zones.

Additional thanks go to the publisher, Springer, for their help and support throughout the editing process and for accepting publishing these proceedings.

Contents

Analysis of Geospatial Data

Assessment on the Distribution and Accessibility to Green Spaces in Mexico's Most Populated Metropolitan Zones

Edali Murillo-Gómez, Marisol Palomar-Ramírez, and Mariana Ramos-Flores

Abstract This project's goal is to make an assessment of the access to urban green spaces in ten of Mexico's most populated metropolitan zones (MZ) by analyzing the relationship between the degree of poverty, social vulnerability, and consequently the access to the benefits these spaces provide that are directly linked to human well-being, as a source of recreational spaces, and places that promote physical and spiritual activities, also known as cultural ecosystem services (FAO 2021). This study consists of three scales of analysis.

- National scale, in which the degree of inequality in the distribution of green areas was assessed in regard to the population according to the Gini Coefficient.
- Metropolitan Zone Level scale, in which in every MZ the geographic location of the poverty and accessibility variables are compared in order to observe whether there is a relationship between green areas and poverty in every AGEB.
- At the local scale an algebra mapping method was employed to obtain the overall quantity of public infrastructure in six parks located in the three most unequal MZ (according to Gini Coefficient). Three are located in AGEBs where the poverty ratio is lower and three in which the ratio is larger, according to CONEVAL. In this case, the accessibility to each park is estimated from mobility and safety costs based on the inclusive public infrastructure that should favor different historically disadvantaged population groups (children, women, the elderly and people with different capabilities) so that they can move across the public space.

E. Murillo-Gómez (✉) · M. Palomar-Ramírez · M. Ramos-Flores
Centro de Investigación en Ciencias de Información Geoespacial, Contoy 137, Col. Lomas de Padierna, Alcaldía Tlalpan, 14240 Ciudad de México, México
e-mail: edalimurillo@gmail.com

M. Palomar-Ramírez
e-mail: mpalomar@centrogeo.edu.mx

M. Ramos-Flores
e-mail: mramos@centrogeo.edu.mx

1 Justification

The rapid growth of metropolitan zones in Mexico have spawned several social and environmental issues that decisionmakers have to face every year with the utmost seriousness (CIDE, LNPP, Centro Mario Molina, IMCO, Citibanamex 2018). The urban population is increasing and there is a global trend of growing cities that extend beyond the limits of their central municipality (Habitat 2020). In this sense, metropolitan areas are defined as a city and its zone of displacement, which consists of suburban, peri-urban and rural areas linked economically and socially, according to the UN Statistical Commission (ibid.).

In order to make cities more inclusive, safe, resilient, and sustainable, Mexico made the commitment to fulfill in 2015 the Sustainable Development Objectives so to attain the goals of the United Nation's Agenda 2030 (United Nations in Mexico 2021). Specifically, Goal 11.7 is a call to action to "provide universal access to green areas, and safe, inclusive and accessible public spaces, particularly for women and children, the elderly and people with different capabilities" (ONU 2015, p. 25).

However, the challenges to ensure everyone access to green areas within the urban space are enormous, because from an environmental justice perspective it is acknowledged that the distribution of natural areas in the built space is quite unequal, and greatly hurts historically disadvantaged peoples (Soja 2016; Bellver Capella 1996).

The concept of environmental justice refers to the equitable distribution of the services provided by urban green areas and their derived benefits (Buckingham & Kulcur 2009). The conceptual framework on environmental justice emphasizes the spatial aspects of the equitable distribution of those services or resources that are valuable to the population and the opportunity for people to use them. In other words, it indicates the access to the inherently urban rights of citizens, including the services that green spaces provide (Soja 2016).

In this sense, the analysis of the access to green areas will be performed from two hypothesis derived from the history of the issue at hand mentioned in the literature and in prior studies about environmental justice.

2 Hypothesis

I.   The relationship between poverty and accessibility → there is a relationship between belonging to a socioeconomic class and access to green areas, since in Mexican cities the poorer zones lack many urban parks.
II. The relationship between poverty and inclusive infrastructure → Access to urban opportunities is conditioned by the urban milieu and the individual's specific autonomy within the space (Mei-Po 1998). Therefore, access to urban green areas is regarded as differentiated for persons from diverse ethnic, gender

and age groups, and green areas in the cities' poor neighborhoods have less inclusive infrastructure.

From these premises, this project seeks to conduct a thorough assessment of the access to urban green spaces in Mexico's most populated metropolitan zones to obtain useful information for decisionmakers so that they can undertake a sustainable socio-ecological-territorial planning as well as public social inclusion policies. In order to achieve this main goal the following secondary objectives have been set:

1. To estimate the degree of environmental justice in Mexico's ten most populated metropolitan zones, in regard to the relationship between the distribution of the poverty levels and the estimation of the accessibility to urban green areas in order to determine whether belonging to an assorted social class determines access, or lack of it, to green spaces.
2. To assess different degrees of environmental justice in the ten metropolitan zones through the estimation of the Gini Coefficient applied to the concentration of access to green areas.
3. To assess the degree of accessibility of two parks in the three most unequal metropolitan zones, according to the Gini Coefficient, through the presence and state of the infrastructure open to the disadvantaged population.
4. To design a new urban green area simulator in one of the most unequal metropolitan zone, located on the poorest AGEBs; so to show the likely growth in the levels of environmental justice.

3 Methodology

3.1 Obtaining Urban Green Areas

To calculate the green areas by MZ, we followed the next procedure: first, the Geostatistical Framework (INEGI 2021) was downloaded, where the green areas were counted by metropolitan zone. It should be noted that only those green areas that comply with the type of small green area present in neighborhoods for daily use were taken. Subsequently, the OSM database was used to complement the database of green spaces taken from INEGI's Geostatistical Framework. In this case, the same criteria were also followed to select the green areas registered in the OSM database. Finally, the information of the two records was verified through the satellite images, and the union of both databases was made.

3.2 Accessibility Per Metropolitan Zone

To analyze metropolitan zones in scale we will focus on Mexico's ten most populated cities, since, due to their size, they suffer most environmental and socioeconomic problems. The selection of metropolitan areas was based upon the National Urban System (CONAPO 2021), which are listed according to population (Table 1)

To study each metropolitan zone, we made a bivariate coroplete map consisting in a comparison between to two layers of categorized information. In this case, variables of interest were poverty and accessibility to green areas, both disaggregated by AGEB, since it was the scale of analysis found for the poverty data.

To estimate the poverty variable, we took the CONEVAL data bases (2015) indicating the percentage of the population in poverty by AGEB. On the other hand, access to green areas was estimated from the 2020 INEGI Population and Housing Census data, and the database created that counts the green areas by MZ. In this case, the method used to estimate access takes the green areas' surface and the distance between the parks' centroids and AGEBs in consideration (Zhang et al. 2011), by applying the following formula:

$$H_p^i = \sum_{j=1}^{k} \frac{A_j}{r_{i,j}^2}$$

where:

H_p^i = accessibility of each AGEB (i)	j = nearest k Parks index
A_j = area of j-nth nearest object	$r_{i,j}$ = distance between AGEB i to object j

Table 1 General data of Mexico's ten most populated metropolitan zones

Metropolitan Zone	Municipalities	Number of inhabitants (2017)	Surface of the urban area km^2	Green areas surface km^2
Valle de México	76	21,650,668	7,866	34.97
Guadalajara	10	4,909,287	3,600	14.153
Monterrey	18	4,603,254	7,657	21.228
Puebla-Tlaxcala	39	3,017,463	2,392	9.328
Toluca	16	2,260,149	2,412	5.157
Tijuana	3	1,996,587	4,423	5.644
León	2	1,743,903	1,760	5.779
Ciudad Juárez	1	1,448,859	3,547	4.240
La Laguna	5	1,342,139	7,889	5.116
Querétaro	5	1,250,429	2,427	4.519

Source Data from (SEDESOL, CONAPO, INEGI, SEDATU, SEGOB 2018) and (CIDE, LNPP, Centro Mario Molina, IMCO, Citibanamex 2018)

3.3   Gini Coefficient

The Gini Coefficient is a measure originally created to estimate income inequality within a country. However, it can be used to appraise any sort of unequal distribution (Longfeng & Seung Kyum 2021). In this case, the Gini method was applied to two variables: the accrued population ratio and the accrued access to green areas, so to assess the degree of inequality within MZ, and then compare these measurements among the ten chosen MZ.

The Gini Coefficient is a number between 0 and 1, in which 0 means perfect equality, namely, that green areas are distributed equally among the population, and, on the other hand, 1 means perfect inequality, which means access to all goods and services by a single individual. Estimation is based upon obtaining the area between a curve of perfect equality and the Lorenz curve; the wider the area between both curves the more unequal society will be, and thus the result will be closer to 1 (Nordhaus 2005).

3.4   Accessibility Per Parks

In this section we pondered on how complicated is that individuals from certain vulnerable groups can visit a park (urban green area) near to home, and can suitably enjoy the outdoor recreation. The infrastructure regarded as ideal for the mobility of these groups is such that ensures the use of space in an autonomous, safe and equal conditions (Gutiérrez Valdivia et al. 2011). So, mobility and population circulation can be unhindered.

In order to estimate optimal routes to access parks, various infrastructures were added through the algebra mapping procedure (ISO4APP 2021). These routes are those streets having an infrastructure suitable for children, women, the elderly and people with different capabilities. Regarding the urban milieu, we consider four road infrastructures, which are: sidewalks, wooden paths, ramps, and streetlamps, besides the shops near the parks. These databases were obtained from the National Housing Catalogue, (INEGI 2021) and the National Statistic Handbook of Economic Units (DENUE), INEGI, respectively.

We explicitly sought the most suitable streets to walk towards the park, namely, the streets with a minimal weigh reflecting a better infrastructure (López et al. 2019). We used the streets located on a 700 m radius from the studied parks, since people will rather walk if the place is located within 400 to 800 m away. Moreover, we considered the route taken by persons with different capabilities. We also took into consideration that if local urban milieus create a suitable environment to replace public and private transportation, then walking becomes a positive and desirable activity (Suárez Lastra & Delgado Campos 2015).

After the algebra mapping, radar graphics were made to compare the parks' infrastructure in order to see if there is any difference between the parks in less poor AGEBs and those located in those where poverty is more widespread, and compare both parks so to look for likely inequalities associated to socioeconomic levels.

Infrastructure thus considered to create the register's matrix by park were: trees, benches, wastebaskets, bicycle ports, courts, shops, walkways, fountains, street-lamps, bus stops, wheelchair ramps, jogging tracks, and restrooms. Additionally, we considered the sidewalks that surround the park.

3.5   Green Areas Simulator

To make the new green areas simulator in a metropolitan zone, and to be able to observe how accessibility measures change, we employed the following method-ology: first, we chose the city on which we would develop the simulator, in this case, the MZ of Querétaro, because its Gini Coefficient, as it would be specified later, is one of the highest in the country; secondly, new parks were digitalized in QGIS, in places within the metropolitan zones identified as unbuilt terrain; finally, the third step was to estimate accessibility by taking into account the actual green areas in the zone as well as the newly added zones.

4   Results

4.1   Accessibility and Poverty in Metropolitan Zones

The results concerning poverty and urban green areas state that in most metropolitan zones poverty is located on the environs, while green areas are concentrated in down-town areas. This relationship determined most of the results obtained in the acces-sibility measurements of urban AGEBs to green areas and in the results of the Gini Coefficient measurements at the metropolitan zone level.

What was found after measuring urban AGEB accessibility in green areas was that most AGEBs with more access are located downtown. On the other hand, those with less access are located on the vicinities. In regard to poverty in metropolitan zones, accessibility is similar, since poorer AGEBs were found mostly on the environs, thus confirming the hypothesis positing that the lesser the poverty in AGEBs the larger the accessibility of urban parks.

We must state that although this behavior was seen in most AGEBs, qualitatively speaking, there were metropolitan zones wherein this situation was more evident: Monterrey, Guadalajara, Puebla-Tlaxcala and Ciudad Juárez. On the other hand, AGEBs showing less were La Laguna, Tijuana, Querétaro and León.

4.2 Ranking of the GINI Coefficient of the Ten Metropolitan Zones

The results of measuring the concentration of accessibility to urban green areas in the ten metropolitan zones are shown in the following Table 2:

In order to compare these results with other visual results obtained by accessibility estimates and the comparison with the geographic location of poverty, we can see that there is a moderately high concordance. Those metropolitan zones where a clear-cut pattern concerning environmental injustice (poorest AGEBs and with less accessibility to green areas) were those in which upon estimating the Gini Coefficient, values were close to 1.

4.3 Accessibility to Parks

Results obtain by algebra mapping confirm that infrastructure surrounding parks is conditioned by the AGEBs' poverty level. If a park was built within a neighborhood with above-average socioeconomic population, this would have a better infrastructure. For example, well lighted streets are safer for women, sidewalks in good condition allow the elderly to walk without fear of falling in potholes, streets with ramps allow people in wheelchairs to go to parks; besides, if streets have trees around the stroll is made pleasant due to the thermal comfort they give, as well as they improve the urban scene (Gutiérrez Valdivia et al. 2011; Suárez Lastra & Delgado Campos, 2015).

On Fig. 1 this is verified, as shown in the results of the Guadalajara MZ regarding two parks. On the left is the map of the Unidad Deportiva Hacienda Real wherein we got most streets marked in red, meaning little accessibility. On the right we see the Rubén Darío park located in a AGEB wherein the poverty percentage is low and there are more streets marked in green, denoting a higher accessibility for the aforementioned park.

Respecting the results of radar graphics, we observed a difference in park infrastructures in less poor AGEBs and parks located in poorer AGEBs. In the following table we can see the overall percentage of the infrastructure within the parks considering the aforementioned variables (tree, benches, wastebaskets, bicycle ports, courts, shops, wooden walkways, fountains, streetlamps, bus stops, jogging tracks, wheelchair ramps, restrooms and the state of sidewalks) (Suárez Lastra & Delgado Campos, 2015). We saw a possible connection between the poverty ratio and how well equipped are these parks for suitable mobilizing the vulnerable population. In the three MZ we see there is an incomplete infrastructure in parks in very poor zones, while their counterparts have better equipped green areas. This relationship is seen in Table 2.

Fig. 1 Algebra maps of infrastructure in two parks in Guadalajara. *Source* INEGI, data from the National Housing Inventory 2016

Table 2 Gini coefficient of ten metropolitan zones

Metropolitan zone	Gini coefficient
Puebla—Tlaxcala	0.7123
Toluca	0.6873
Guadalajara	0.6859
Querétaro	0.6802
La Laguna	0.6296
León	0.6170
Valle de México	0.6124
Tijuana	0.4945
Monterrey	0.4672
Ciudad Juárez	0.4367

Source Data from INEGI and OpenStreetMap

So, we can conclude that urban green areas located in poor neighborhoods get a less money, which diminishes the quality of the leisure this population can enjoy (Fisher et al. 2018) (Table 3).

4.4 Green Areas Simulator

The results obtained through the green areas simulator in Querétaro show that increasing the number of those areas in those poverty-stricken AGEBs will also

Table 3 Infrastructure percentage in six parks

Metropolitan zone	Parks	% poverty	% of infrastructure
Puebla—Tlaxcala	Zócalo de Huejotzingo	High	79
	Benito Juárez	Low	83
Guadalajara	Hacienda Real	High	67
	Rubén Darío	Low	76
La Laguna	Plaza el Triángulo	Very high	62
	Lienzo Charro	Low	71
Poverty ratio (CONEVAL): Low	Medium	High	Very high

Source Data from INEGI and Google Maps

increase the accessibility and environmental justice measurements. As we can see in Fig. 2, the AGEBs selected are located mostly on the environs and on those with higher poverty ratios.

Besides qualitatively improving accessibility and environmental justice, this measure also show improvements in the Gini estimates. In the past, this metropolitan zone showed a Gini Coefficient of 0.6802, and after implementing the green areas simulator the resulting coefficient value was 0.6770.

5 Discussion

Derived from the results, we show that environmental injustice in Mexican metropolitan zones is a socioenvironmental issue that impact differently among diverse age groups, genders, capabilities and social classes. This could be proven in the results obtained by our study. Initially, we qualitatively verified the urban green parks concentration, and thus their accessibility, in those less poor AGEBs. Later, we corroborated the environmental injustice through the Gini Coefficient, since it quantitively reflected that none of the metropolitan zones had an equal access distribution. Finally, by obtaining infrastructure maps we proved that the routes showed the necessary infrastructure to ensure the access of the vulnerable population is highly limited.

Thus, the environmental injustice problem, given its complex nature, must be faced by the government authorities as a priority, especially for its impact upon the quality of life of its inhabitants. Furthermore, we must consider that cities have high growth rates, and so it is urgent to develop a wide-ranging public urban planning policy sufficiently funded so to revert environmental injustice throughout the entire metropolitan zone.

Moreover, we recommend implementing wide-ranging public policies with an intersectional outlook, which would encompass the considerations and adaptations needed for any public policy that aims at eradicating social and environmental inequality with no harmful effects to the historically disadvantaged populations.

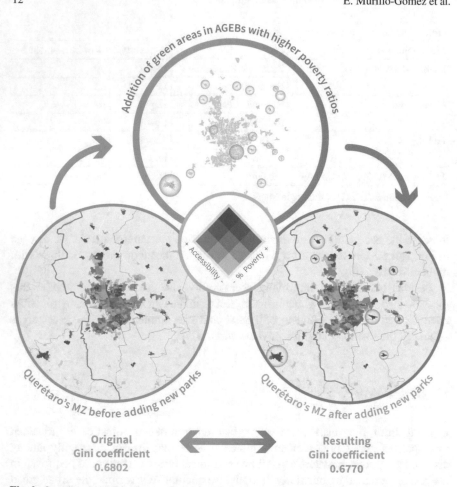

Fig. 2 Querétaro's MZ before and after adding new parks. *Source* Data from INEGI, OpenStreetMap and CONEVAL

6 Conclusion

As limitations, it was obtained that the results on the degree of environmental justice, measured through the access that populations have to green areas, should be considering that there was a lack of information on the number of green areas present in metropolitan zones. Therefore, it is considered important for government organizations to have an inventory, which must be updated periodically, with classifications of the type of green area found in the territory.

In the same way, the scale of analysis is considered as a limitation of the investigation because, due to the type of problem analyzed, it is important to consider the information at the block level. Thus, it is necessary for government agencies to make available information on poverty and green areas (counting vegetation, state of it, and presence and state of infrastructure) at the block level.

Regarding the results of the GINI, it must be considered that it is an aggregate measure that does not necessarily adhere to the conditions of environmental justice present in metropolitan zones. Similarly, the results of the green areas simulator show that the public policy of increasing green areas in the most marginalized areas of the MZ does not necessarily represent a significant change in the presence or absence of environmental justice. For this reason, two actions must be taken to resolve these limitations: first, the construction of other complementary indicators that will allow knowing in a general way the degree of environmental justice present in metropolitan zones is considered necessary. Second, it is necessary to consider that public policies in charge of promoting urban planning based on environmental justice must consider the different aspects of the problem: discrimination, poverty, gender inequality, gender violence, pollution, and the lack of green infrastructure in the MZ.

Regarding upcoming research plans, we recognize the need to create more far-ranging and closer to reality measures concerning human mobility, particularly for vulnerable groups. These measurements and the availability of data will allow us to get more information on the accessibility associated to each city's special context, as well as to design better urban development plans and to implement public social inclusion policies so to make the public space in metropolitan zones fairer for all.

References

Capella VB (1996) El movimiento por la justicia ambiental: entre el ecologismo and los derechos humanos. Anuario de Filosofía del Derecho XIII, 327–347

Buckingham S, Kulcur R (2009) Gendered geographies of environmental injustice. Antipode 41:659–683

CIDE, LNPP, Centro Mario Molina, IMCO, Citibanamex (2018) Índice de Ciudades Sostenibles 2018: Desafíos, rumbo al 2030, de los Objetivos de Desarrollo Sostenible en las zonas metropolitanas de México. Reporte Completo. Banco Nacional de México, S.A. http://www.lnpp.cide.edu/indicedeciudadessostenibles2018

CONAPO (2021) Consejo Nacional de Población. http://conapo.gob.mx/

CONEVAL (2015) Consejo Nacional de Evaluación de la Política de Desarrollo Social. https://www.coneval.org.mx/

FAO (agosto de 2021) Servicios ecosistémicos y biodiversidad. http://www.fao.org/ecosystem-services-biodiversity/es/

Fisher D, Hamstead Z, Ilieva RT, Wood S, McPhearson T (2018) Geolocated social media as a rapid indicator of park visitation and equitable park access. Comput Environ Urban Syst 38–50

Gutiérrez Valdivia B, Fonseca M, Casanovas R, Ciocoletto A, Muxí Martínez Z (2011) Qué aporta la perspectiva de género al urbanismo. Feminismos 105–129

INEGI (2021) Instituto Nacional de Estadística and Geografía. https://www.inegi.org.mx/

ISO4APP (agosto de 2021). http://www.iso4app.com/

Longfeng W, Seung Kyum K (2021) Exploring the equality of accessing urban green spaces: a comparative study of 341 Chinese cities. ScienceDirect 1–9

López Ortega J, López-Sauceda J, Carrillo J, Sandoval J (2019) Método de construcción de dígrafos a partir de redes viales reales en mapas digitales con aplicaciones en la búsqueda de rutas óptimas. Informes de la Construcción

Kum MP (1998) Space–time and integral measures of individual accessibility: a comparative analysis using a point-based framework. GeoGraphical Anal 191–216

Nordhaus S (2005) Economía. McGraw-Hill, Ciudad de México

Habitat ONU (2020) Estado Global de las Metrópolis 2020 - Folleto de Datos Poblacionales. Programa de las Naciones Unidas para los Asentamientos Humanos. https://unhabitat.org/es/node/144450

ONU (2015) Transformar nuestro mundo: la Agenda 2030 para el Desarrollo Sostenible. Asamblea General. https://www.unfpa.org/sites/default/files/resource-pdf/Resolution_A_RES_70_1_SP.pdf

SEDESOL, CONAPO, INEGI, SEDATU, SEGOB (2018) Delimitación de las zonas metropolitanas de México 2015. https://www.gob.mx/conapo/documentos/delimitacion-de-las-zonas-metropolitanas-de-mexico-2015

Suárez Lastra M, Delgado Campos G (2015) Entre mi casa y mi destino. Movilidad y transporte en México. Universidad Nacional Autónoma de México, México

Soja EW (2016) La ciudad y la justicia espacial. Justicia e injusticias espaciales, pp 99–106

United Nations in Mexico (2021) Results Report 2020. https://www.onu.org.mx/wp-content/uploads/2021/09/UN-REPORT2020Links.pdf

Zhang X, Lu H, Holt J (2011) Modeling spatial accessibility to parks: a national study. Int J Health Geogr

Geomatics Assessment of Water Resources in a Transboundary Basin

Violeta Yoalli Alvarado-Arriaga, Felipe Omar Tapia-Silva, and Fabiola Sagrario Sosa-Rodríguez

Abstract Usumacinta is a transboundary basin located between Mexico and Guatemala. Human activities in the last decades have impacted this basin. Consequently, it has impacted water resources also. This research was implemented the Driver-Pressure-State-Impact-Response framework (DPSIR), and it has been enriched with the geomatics approach to figure out the main factors that alter water resources and optimize the management strategies. However, this article only talks about *Drivers-Pressures-State* directly related to water availability. Drivers are the water requirements for urban-public use, agricultural use, industrial use, and other uses. Pressures are changes in surface water bodies and increased agricultural frontier. Finally, indicators of *State* are water balance and water availability. The analysis results show that even though the Usumacinta River is one of the largest rivers in Mexico and Central America, the municipalities within the basin have a deficit in their local water availability.

Keywords Transboundary basin · Water availability · Water balance · Consumption of water

1 Introduction

Water is a fundamental resource for population well-being, whose comprehensive management is essential to sustainable development achieve. For this, water use must not compromise the ecosystems' vitality and availability. However, water management requires a collaborative effort in a transboundary basin (Babel and Eiman

V. Y. Alvarado-Arriaga (✉)
Doctorate Program in Energy and Environment, Metropolitan Autonomous University, Iztapalapa, Mexico City, Mexico

F. O. Tapia-Silva
Department of Hydrology, Metropolitan Autonomous University, Iztapalapa, Mexico City, Mexico

F. S. Sosa-Rodríguez
Department of Economics, Metropolitan Autonomous University, Azcapotzalco, Mexico City, Mexico

© The Author(s), under exclusive license to Springer Nature Switzerland AG 2022
R. Tapia-McClung et al. (eds.), *Advances in Geospatial Data Science*, Lecture Notes in Geoinformation and Cartography, https://doi.org/10.1007/978-3-030-98096-2_2

15

2012) and the development of a common management framework. Thus, countries must establish transparent cooperation and communication strategies; otherwise, there is a risk that decision-making will be confusing and complex (Lorenz et al. 2001; Yuan et al. 2020). The main obstacles for cooperation include differences in visions and hydro-political strategies, development priorities, economic situation, social stratification, legislation, insufficient data, ineffective information exchange, and lack of monitoring protocols (Skoulikaris and Zafirakou 2019). Geomatics is one of the disciplines with the most significant capacity to generate, analyze, homogenize and democratize information for accurate assessments that help sustainable water management; therefore, it is the best discipline to address the challenges above (Tapia-Silva 2014).

Geomatics has been widely used to assess water resources, particularly where there is no infrastructure to water bodies monitoring or temporal data is not available. This discipline can generate information on water vulnerability and contamination, surface water modeling, or water balance modeling (Chen et al. 2015; Molina et al. 2014; Tapia 2014). This discipline is rapidly evolving in instrumentation, sensor integration, automation, and analysis methods (Molina et al. 2014), thus providing high-quality information produced at different spatial and temporal scales, making them versatile accessible to different types of analysis (Bhaduri et al. 2016). In addition, geomatics provides an integrative framework that supports timely and effective decision-making and fosters collaboration and consensus-building (Greg and Abbas 2017). Nevertheless, it requires conceptual frameworks that structure the dynamics of the socio-environmental system associated with water to analyze, understand and explain its complexity. The framework selected for the present work was the DPSIR (Driver-Pressure-State-Impact-Response), which allows precise and straightforward visualization of the complex cause-effect relationships between human activities and their interaction with the environment (Lewison et al. 2016). The model DPSIR is comprised of five components. The first component, *driving* force, refers to a phenomenon that governs ecosystem change and could be either anthropogenic or natural. Anthropogenic factors may be related to food, clean water, employment, or energy, while natural driving forces include earthquakes, volcanic eruptions, or other natural phenomena. The second component, *pressures*, results from mechanisms initiated by driving forces that can affect ecosystems by altering their condition. The third component, *State,* results from pressures that affect the environmental quality of air, water, soil and results from the combination of physical, chemical, or biological conditions. The fourth component, *Impacts,* refers to changes in the ecosystem's condition that affect their functioning and generate economic, social, or health problems. Finally, the fifth component, *Responses,* relates to public policies or community measures that help avoid, mitigate or restore impacts (Kristensen 2004; Oesterwind et al. 2016). However, there is a risk of oversimplifying water problems if the multiple interactions of the different components are not considered; addressing this weakness requires a deep understanding of the components (Patricio et al. 2016).

Geomatics and the DPSIR model have already been used to assess water resources. For example, Caeiro et al. (2004) proposed a management system that reconciles development conflicts with the conservation of an estuary in Portugal. After identifying the indicators to be evaluated, management strategy areas were determined, and environmental quality was assessed. Mattas et al. (2014) identified the main factors affecting the water quality resources and the water reserve decrease and made a soil mapping to propose policy strategies. Lalande et al. (2014) constructed an indicators system to assess water quality, land uses, and their influence on the ecological quality of rivers at three spatial scales. One of the most recent contributions is that of Nguyen et al. (2020), who have evaluated water vulnerability due to climate change, estimating exposure indices sensitivity, adaptive capacity, water vulnerability, and distribution of indicators at the local and regional scales. In all these works, the DPSIR model helped improve the relationship understanding between water and anthropogenic pressures, identifying the best strategies to improve its management, and supporting some indicators.

Thus, the current study is focused on making a diagnosis and determining the causal relationships which affect surface water availability in a transboundary basin through the use of geomatic and the first three components of the DPSIR framework (i.e., *Driving forces-Pressures-State*). This first phase describes how the variables that have determined the changes in water resources are linked spatiotemporally, allowing a first diagnosis given the lack of studies with this approach in the area. The analysis does not include the *Impact* and *Response* components since participatory methods will identify the responses implemented to address the current problems.

The most relevant contributions of the research are:

1. Develop a methodology that integrates the geomatics virtues and the DPSIR framework to develop such a diagnosis,
2. Discuss and highlight the collaboration importance of the involved countries in a transboundary basin to generate helpful information for collaborative and integrated management,
3. Identify indicators that help to systematically measure water resources and the factors that significantly impact them.

The interrelationship between the *Driving forces* will be evaluated considering the population's basic needs and the basin's activities for different water uses (agricultural, industrial, public-urban, and others). The *Pressure* components show the ecosystems changes due to the demand for resources to satisfy the population demand to assess variations on surface water bodies and the increase in water demand. Finally, the *State* components measure how the resources are affected by pressures exerted on them, such as the water balance and water availability at the basin and municipal level. Since this is a transboundary basin, the lack of available information restricts some indicators evaluation to only one of the two countries. Accurately, the indicators of water requirements, cropland increased, and water availability was estimated only for the Mexican area of the Usumacinta basin. Meanwhile, indicators such as changes in surface water and water balance were calculated for the entire basin.

Fig. 1 Usumacinta Basin

2 Methodology

2.1 Study Area

The Usumacinta basin (Fig. 1) is transboundary. Mexico accounts for 44.2% of its territory, Guatemala for 55.7%, and Belize for 0.04%. The coordinates bound it 92°36'53.1 "W and -89°2'38.3 "W longitude and 18°32'38.4 "N and 14°51'10.8 "N latitude. Because of its location and extension, the Usumacinta basin is home to 66% of Mexico's biological diversity in a large mosaic of ecosystems, including forests, jungles, swamps, and mangroves (March-Mifsut and Castro 2010). This region is home to more than 5.3 million inhabitants, one–third of whom are indigenous peoples living in a highly marginalized situation (López 2020). This situation makes it difficult for conservation promotion since the priority for these groups is to survive, for which they exploit the ecosystems in which they live in an unsustainable manner.

The Usumacinta basin is a complex region, so a DPSIR framework was constructed considering social, economic, and environmental factors that, in turn, made the structure of the complexity of the system involved of the water resources. However, the methodology and results presented below are only directly related to water availability in the basin.

First, the DPSIR conceptual framework is shown (Fig. 2), constructed based on bibliographic research. The indicators selection followed this criterion:

1. Those were representative and frequent.
2. That their evaluation was feasible from geomatics.
3. That the indicators were environmental, economic, and social in order to ensure a sustainable approach.

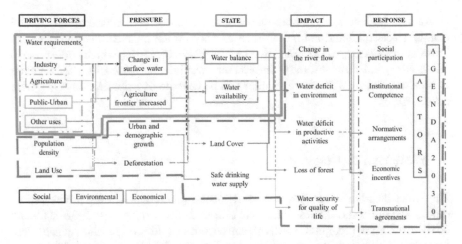

Fig. 2 Driving Forces, Pressures, and States considered assessing water sustainability in the Usumacinta basin. In this framework, three types of components are considered: Social (red), economic (yellow), and environmental (green). Components not explained in this publication (gray)

3 Driving Forces Evaluation

The study area was delimited from a hydrologic model for the definition of surface water connectivity, implemented in GRASS GIS 7.8, consisting of the calculation of flow directions and accumulations and the use of a threshold value of accumulations. This model will determine streams and catchment areas immersed in the Usumacinta basin, based on images from the Digital Elevation Model (DEM) performed by NASA's Shuttle Radar Topography Mission (SRTM). Once it was done, the administrative entities circumscribed in the basin were demarcated.

From databases consulted by the National Water Commission (CONAGUA 2020), the demand for surface water for different uses (agriculture, industry, public-urban, and other uses) was evaluated, and its distribution by the municipality from 1996 to 2019.

4 Pressure Evaluation

The variation of surface water, decade by decade, in the Usumacinta basin was obtained from the maps generated by Pekel et al. (2016), which show the surface water distribution over time. The information was processed with the Google Earth Engine, analyzing the presence or absence of water between 1990 and 1999, 2000–2009, and 2010–2019, comparing pixel by pixel the presence of persistent and seasonal water bodies of the Usumacinta basin.

Another indicator of pressure on water resources is the increase of agricultural frontier since soil deterioration can have effects on water balance. The source of this information was maps performed by SIAP (2012, 2015, 2017), which correspond to 2012, 2015, and 2017. From the information provided by these maps, the increase of area covered by agriculture frontier and percentage of irrigation modality were estimated.

5 State Evaluation

The water balance is composed of 4 elements: precipitation, evapotranspiration, runoff, and infiltration. These data were estimated for two periods: 2002–2009 and 2010–2017. First, precipitation was obtained from a CHIRPS algorithm, which arises from the satellite data and weather station integration (Funk et al. 2015). The annual precipitation was added, and the annual average between periods was estimated and interpolated by b-splines. Subsequently, evapotranspiration was obtained from the PML_V2 product (Pekel et al. 2016), resulting from the MODIS conjunction and GLDAS products. This information was used to obtain the terrestrial evapotranspiration, which includes vegetation transpiration, soil evaporation, evaporation of water intercepted in the canopy, and evaporation from water bodies) (Zhang et al. 2019). The annual average evapotranspiration of 2002–2009 and 2010–2017 were estimated and interpolated by b-splines for the annual evapotranspiration.

On the other hand, the runoff was obtained following the indirect method suggested in the NOM-011-CNA-2000. It is necessary to calculate the K value, which is the result of subdividing the basin into homogeneous zones according to type and land use and obtaining the weighted average of all of them. In turn, the land cover maps were obtained from the LANDSAT5 and OLI8 images classification using random forest and validated according to the method suggested by Olofsson et al. (2014) and obtaining values relative to the area (Pontius and Millones 2011) around 80%. From this value, the runoff coefficient is estimated according to the following equation

$$Ce = K (P - 250) / 2000 + (K - 0.15) / 1.5 \qquad (1)$$

where Ce is the runoff coefficient, K factor which was weighted previously, and P is the precipitation. Finally, infiltration was obtained from the mass water balance, which means.

$$I = P - EVT - Es \qquad (2)$$

where I is infiltration, EVT is evapotranspiration, and Es is runoff.

Finally, the annual surface water availability average was estimated according to the methodology suggested in NOM-011-CNA-2000, which suggests that:

Annual availability average $=$ annual runoff average

$-$ annual committed volume average.   (3)

The runoff is the sum of the runoff value from each pixel, and the average annual volume committed is the sum of the volume granted. The standard suggests that this measurement should be made with downstream water volumes; this estimate was made at the municipal level by performing zonal operations.

6   Results

The indicators to diagnose water availability in the Usumacinta basin only consider *Driving forces-Pressures-State* components. In this framework, three types of components are considered: social (red), economic (yellow), and environmental (green). Components whose estimation is in progress are shown without color. The DPSIR framework described results from selecting indicators providing evidence of the complexity of the water resources systems (Fig. 2).

Water requirements for public-urban, industrial, agriculture uses, among others, are the primary surface water demand source in the basin, so that, as this resource needs increasing to cover other basic needs like food, hygiene, productivity, and services, the forces that pressure the environment, and specifically the water, are then increased. In this sense, the most significant water demand in the Mexican Usumacinta basin territory from 1996 to 2019 has been agriculture, which consumed 45% of the concessioned water volume up to 2019, followed by other uses (a category provided by CONAGUA without specifying what kind). Other uses have consumed 36% of water, especially in Carmen and Tenosique municipalities; use is public-urban, which has been allocated only 19% of the surface water granted, despite being the use with the best distribution. It is important to note that there are municipalities where water is not used for this use, such as Carmen, Palizada, or La Libertad. Finally, the industry has been granted only 0.2% of the surface water.

Furthermore, Fig. 3 shows that the states with the highest consumption are Carmen, Palizada, Balancán, and Ocosingo. The inequitable distribution of water can mean present problems, which may increase in the future, potentially generating conflicts between the actors involved to guarantee its access. The preceding, although the study area is considered to have greater water availability in the country.

On the other hand, the location and persistence of bodies of water are affected by climatic factors and human activities. However, in Fig. 4, the temporal evaluation of this indicator did not show changes related to a direct impact due to the extraction of permanent surface water, but there was a slight increase in seasonal water bodies, which in part can be explained by an increase in precipitation as seen in Fig. 6.

As explained at the beginning of the results, the committed water volume for agricultural use is the largest territory of the Usumacinta basin in Mexico. This situation can be explained by Fig. 5, which shows the rapid increase of the agricultural

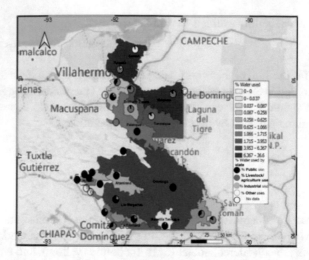

Fig. 3 Map of water consumption for agricultural, public–urban, industrial and other users in the Usumacinta basin (Mexico)

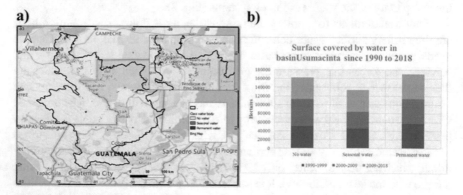

Fig. 4 a Surface water bodies in the Usumacinta basin. **b** Temporal comparison of permanent and seasonal water bodies

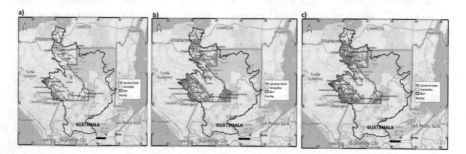

Fig. 5   Increase of the agricultural frontier in the Usumacinta basin (Mexico)

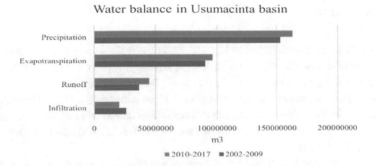

Fig. 6 Comparison of the elements of the water balance in the Usumacinta basin

frontier from 171,082 to 256,510 hectares, from 2012 to 2017, of which only 1.5% is irrigated agriculture (SIAP 2012; SIAP 2017). Interestingly, the increase in agricultural coverage mainly focuses on municipalities where agricultural concessions are located (Fig. 3), despite a small percentage being registered as irrigated agriculture. This result may indicate an increase in the use of water for irrigation in an unauthorized way.

On the other hand, the water balance is an environmental variation and changes indicator in the landscape. Figure 6 shows a pixel-by-pixel estimation summary at the basin level for each of the water balance components. This Figure showed an increase in precipitation, runoff, and evapotranspiration for the last decade but a reduction in infiltration. This result may be a consequence of land-use change, as discussed in the analysis.

Finally, although water availability directly depends on the water balance, it can be affected by the required demand for water to satisfy basic needs, as explained above. Figure 7 shows water availability in the Usumacinta basin Mexican municipalities. Of the 29 municipalities, 21 showed a deficit, with Carmen, Palizada, Balancán, and Emiliano Zapata being some of the highest deficits in the lower basin. Furthermore, the deficit increased in most municipalities during the last two decades.

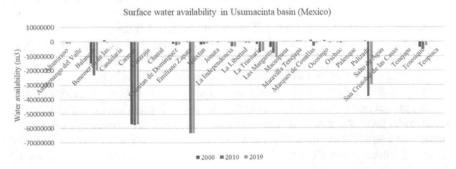

Fig. 7 Water availability by municipality in the years 2000, 2010 and 2019 in the Usumacinta basin

7 Discussion

The DPSIR framework shown in Fig. 2 results from the systematic weighting and structuring of water resources factors. Only *Driving forces* and *Pressures* that affect water availability are analyzed since this first phase describes how the cause-effect relationships of the analyzed factors impact surface water availability. The indicators obtained can help to establish spatially differentiated management and administration of water resources. Each of the components of the framework and their interaction will be discussed below.

First, it is necessary to highlight that one of the obstacles to doing a water analysis and management at the basin level is the lack of sufficient spatio-temporal information and binational agreements to generate it.

Figure 3 shows the lower part of the basin municipalities being the ones that consume the highest amount of water, notably Carmen, Palizada, and Balancán, followed by Ocosingo, Las Margaritas, and Emiliano Zapata due to the increase in the agricultural frontier; therefore, being the agricultural use the one that has consumed 45% of the concessioned water from 1996 to 2019. On the other hand, for at least half of the municipalities analyzed, the most important consumption of water is destined for public-urban use, like, for example, Ocosingo, Palenque, or Altamirano, to name a few. This imbalance in water supply implies competition between users, and possibly in low rainfall season, it could trigger runoffs, lack of access to this resource, and even conflicts. However, it is essential to mention that there are no relevant records of water intended for industry, which is a sign of a lack of economic development in the region.

Due to the extraction volumes from surface waters, it could be expected that the covered surface water bodies' area would be altered, but the pressure analysis of the surface water changes indicator gave no evidence of changes due to human activities. However, environmental variations were recorded, such as a slight increase in the covered areas by seasonal water bodies in the last decade. This result is, as shown in image 6, precipitation also increased during this period.

On the other hand, the agricultural areas increase may be a cause of altering important basin environmental factors, such as the soil infiltration capacity, which significantly decreases when forest soil is replaced by agricultural soil since the soil structure is more granular in forests. The change from forest cover to agriculture can also increase surface runoff because there are no obstacles to stop runoff and give time to soil to infiltrate the water. In this scenario, the increase in the agricultural frontier (Fig. 5) is essential as pressure on the water resources indicator.

In this sense, the comparison shown in image 6 points out that: in the 2002–2009 period, precipitation was lower than in the 2010–2017 period, which means in the last decade, there was more water available for plants and soil. The image shows that runoff and evapotranspiration were higher in the last decade than in the previous one. However, infiltration did not correspond to expected, as it was lower in 2010–2017 than in the previous decade, despite higher precipitation. This can be explained due to the change in land use and, consequently, the change in soil permeability, so the volume of water infiltrated reduced.

Despite the above, results regarding the availability of water in the Usumacinta basin are alarming, because although the water balance showed that in the period 2010–2017, there was more significant runoff and, therefore, more surface water available, the basin's demand for water exceeds the available surface water (runoff). This can be seen in Fig. 7, where only 8 of the 29 municipalities have a slight water surplus, while the rest show a deficit. This means that in municipalities where local runoff is not enough to satisfy their water needs, therefore, the management of this resource must be replaced so that access to water is sustainable and safe. On the other hand, Fig. 7 shows four alarming cases, which are the municipalities of Carmen, Balancán, Palizada, or Emiliano Zapata, whose water deficit is alarming. This means that the local runoff is far from satisfying the water requirements, but fortunately, these municipalities are in the lower basin, so it is likely that the main channel of the basin can supply water to these municipalities, although local runoff does not allow self-sufficiency. However, this does not imply that it is not necessary to optimize water management in the municipalities above.

8 Conclusion

The DPSIR framework and geomatics tools provided a spatial approach that contributes to the comprehensive analysis of the leading environmental and socioeconomic drivers and pressures acting on water resources at the basin and municipal scale. Both tools made it possible to understand and systematize complex information and evaluate it without a spatial and temporal perspective. Moreover, with the proposed methodology, it is possible to implement it in other adjacent basins or repeat the study in the future to improve the diagnosis.

Although geomatics virtues and its enrichment with the DPSIR framework, unfortunately, many indicators could not be evaluated at the basin level because Guatemala has no information about how water is used. This missing information is a clear sign of the cooperation lack and agreements between Mexico and Guatemala to generate information that could allow adequate management of the shared basins.

Finally, the evaluation of the results shows the urgency for developing strategies for water use optimization in the Usumacinta basin because even though it has one of the greatest rivers, this abundance is not reflected in the adequate water resources management at the municipal level. It should be noted that when referring to water resource management, we are not only calling for better water management but also for the terrestrial ecosystems care and conservation, which will allow the environment to provide an adequate water balance, i.e., that runoff and infiltration are conducive to increased water availability.

References

Babel M, Eiman K (2012) A global analysis of river basins science and transboundary management. UNU-INWEH

Bhaduri A, Bogardi J, Siddiqi A, Voigt H, Vorosmarty C, Pahl-Wostl C, Osuna V (2016) Achieving sustainable development goals from a water perspective. Front Environ Sci

Caeiro S, Mourão I, Costa MH, Painho M, Ramos TB, Sousa S (2004) Application of the DPSIR model to the Sado Estuary in a GIS context – Social and Economical Pressures. In: 7th AGILE conference on geographic information science" 29 April-1 May 2004, Heraklion, Greece

Chen Y, Liu R, Barrett D, Gao L, Zhou M, Renzullo L, Emelyanova I (2015) A spatial assessment framework for evaluating flood risk under extreme climates. Sci Total Environ 15(538):512–523. https://doi.org/10.1016/j.scitotenv.2015.08.094

CONAGUA (2020) Consulta a la base de datos del REPDA. Obtenido de https://app.conagua.gob.mx/consultarepda.aspx

Funk C, Peterson P, Landsfeld M, Pedreros D, Verdin J, Shukla S, Michaelsen J (2015) The climate hazards infrared precipitation with stations—a new environmental record for monitoring extremes. Nature 2(150066). https://doi.org/10.1038/sdata.2015.66

Greg S, Abbas R (2017) Sustainable development and geospatial information: a strategic framework for integrating a global policy agenda into national geospatial capabilities. Geo-Spat Inf Sci 20(2):59–76

Kristensen P (2004) DPSIR Framework. a comprehensive/detailed assessment of the vulnerability of water resources to environmental change in Africa using river basin approach. Nairobi, Kenya: UNEP Headquarters

Lalande N, Cernesson F, Decherf A, Tournoud M-G (2014) Implementing the DPSIR framework to link water quality of rivers to land use: methodological issues and preliminary field test. Int J River Basin Manag 12(3):201–217

Lewison R, Rudd M, Al-Hayek W, Baldwin C, Beger M, Lieske S, Hines E (2016) How the DPSIR framework can be used for structuring problems and facilitating empirical research in coastal systems. Environ Sci Policy 110–119

López P (27 de Agosto de 2020) Avanza el deterioro de la Cuenca del Usumacinta. Gaceta UNAM

Lorenz CM, Gilbert AJ, Vellinga P (2001) Sustainable management of transboundary river basins: a line of reasoning. Regional Environ Change 2:38–53. https://doi.org/10.1007/s101130100023

March-Mifsut I, Castro M (2010) La cuenca del río Usumacinta: Perfil y perspectiva para su conservación y desarrollo sostenble. En H. Cotler, *Las Cuencas Hidrográficas de México. Diagnóstico y Priorización*. Ciudad de México: SEMARNAT, pp 193–197

Mattas C, Voudouris K, Panagopoulos A (2014) Integrated groundwater resources management using the DPSIR approach in a GIS environment: a case study from the Gallikos river basin, North Greece. Water 1043–1068

Molina JL, Rodríguez-Gonzálvez P, Molina MC, González-Aguilera D (2014) Geomatic Methods at the service of water resources modelling. J Hydrol 509,150–162. "http://dx.doi.org/https://doi.org/10.1016/j.jhydrol.2013.11.034" \t "_blank" https://doi.org/10.1016/j.jhydrol.2013.11.034

Nguyen T, Ngo H, Guo W, Nguyen HQ, Luu C, Dang KB, Liu Y, Zhang X (2020) New approach of water quantity vulnerability assessment using satellite images and GIS-based model: an application to a case study in Vietnam. Sci Total Environ 237. https://doi.uam.elogim.com/https://doi.org/10.1016/j.scitotenv.2020.139784

Oesterwind D, Rau A, Zaiku A (2016) Drivers and pressures e Untangling the terms commonly used in marine science and policy. J Environ Manage 181:8–15

Olofsson P, Foody GM, Herold M, Stehman SV, Woodcock CE, Wulder MA (2014) Good practices for estimating area and assessing accuracy of land change. Remote Sens Environ 148:42–57. https://doi.org/10.1016/j.rse.2014.02.015

Patricio J, Elliot M, Masik K, Papadopoulou K, Smith C (2016) DPSIR—two decades of trying to develop a unifying framework for marine environmental management? Front Mar Sci 3. https://doi.org/10.3389/fmars.2016.00177

Pekel JF, Cottam A, Gorelick N et al (2016) High-resolution mapping of global surface water and its long-term changes. Nature 540:418–422. https://doi.org/10.1038/nature20584

Pontius Jr RG, Millones M (2011) Death to Kappa: birth of quantity disagreement and allocation disagreement for accuracy assessment. Int J Remote Sens 32(15):4407–4429. https://doi.org/10.1080/01431161.2011.552923.DOI:10.1080/01431161.2011.552923

SIAP (2015) Conjunto de datos vectoriales de la frontera agrícola de Mexico, Serie II. Ciudad de México. Obtenido de. http://infosiap.siap.gob.mx/gobmx/datosAbiertos.php

SIAP (21 de 06 de 2017) Conjunto de datos vectoriales de la frontera agrícola de Mexico, Serie III. Ciudad de México

SIAP (30 de 11 de 2012) Conjunto de datos vectoriales de la frontera agrícola de Mexico, Serie I. Ciudad de México

Skoulikaris C, Zafirakou A (2019) River basin management plans as a tool for sustainable transboundary river basins' management. Environ Sci Pollut Res 26:14835–14848. https://doi.org/10.1007/s11356-019-04122-4

Tapia-Silva FO (2014) Avances en geomática para la resolución de la problemática del agua en México. Tecnología y Ciencias Del Agua 5(2):131–148

Yuan L, He W, Degefu DM, Liao Z, Wu X, An M, Zhang Z, Ramsey TS (2020) Transboundary water sharing problem; a theoretical analysis using evolutionary game and system dynamics. J Hydrol 582. https://doi.org/10.1016/j.jhydrol.2019.124521

Zhang Y, Kong D, Gan R, Chiew F, McVicar T, Zhang Q, Yang Y (2019) Coupled estimation of 500 m and 8-day resolution global evapotranspiration and gross primary production in 2002–2017. Remote Sens Environ 222:165–182. https://doi.org/10.1016/j.rse.2018.12.031

Geospatial Analysis of Clandestine Graves in Baja California: New Approaches for the Search of Missing Persons in Mexico

José L. Silván-Cárdenas, Ana J. Alegre-Mondragón, and Jorge Ruiz-Reyes

Abstract Clandestine graves are an expression of extreme violence whose findings have increased in Mexico in recent years. In this paper we use two concepts already studied: spatial clustering and clandestine space and integrate them into a spatial model within a web application, with the aim of improving the previous model by reducing the areas with the highest probability of finding more clandestine graves in Baja California (BC), based on information from 52 grave points already located by the local attorney general's office. The results confirm that, by incorporating the analysis of point patterns, the search area is substantially reduced ($<10\%$). The model ensures that the final search areas will be within practical distances from most urban settlements, 39 min in the case of BC.

1 Introduction

Clandestine grave(s) is a generic term to refer to places where perpetrators of crimes deposit human remains of murdered people in order to conceal any evidence (Komar 2008). Most of the time, the families of these victims may not even know their relatives are deceased and may still consider them missing persons. If known to be deceased, the families have been denied the opportunity for their customary mortuary rituals, closure, and receipt of death certificates, which are necessary in order to move forward with many legal matters.

J. L. Silván-Cárdenas (✉) · A. J. Alegre-Mondragón
Centro de Investigación en Ciencias de Información Geoespacial, Contoy 137, Col. Lomas de
Padierna, Alcaldía Tlalpan, 14240 Ciudad de México, México
e-mail: jsilvan@centrogeo.edu.mx

A. J. Alegre-Mondragón
e-mail: jalegre@centrogeo.edu.mx

J. Ruiz-Reyes
Universidad Iberoamericana, Prolongación Paseo de la Reforma 880,
01219 Mexico City, Mexico
e-mail: jorge.ruiz@ibero.mx

© The Author(s), under exclusive license to Springer Nature Switzerland AG 2022
R. Tapia-McClung et al. (eds.), *Advances in Geospatial Data Science*, Lecture Notes
in Geoinformation and Cartography, https://doi.org/10.1007/978-3-030-98096-2_3

In Mexico, the number of disappeared people has grown exponentially since the year 2006, when the government openly declared the war against drug cartels, and the search for missing persons in the country turned undesirably into the search for clandestine graves. The discovery of clandestine graves was driven by organized groups of family members of missing persons together with nongovernmental organizations which developed their own rudimentary search methods. Over the years, these efforts resulted both in the discovery of a considerable number of clandestine graves across the country and the creation of a national and local commissions for search of missing persons, as well as the creation of special laws that recognize the search activity and obligates Government to maintain records of the missing persons.

As data is being accumulated, more sophisticated search methods are enabled. The use of geographic information systems for the analysis of point patterns have been used to search for clandestine graves in different places and contexts with armed conflicts or extended violence. These methods coincide and emphasize the quality of the data, the social and cultural context, as well as the care in the elaboration of models, since they are abstractions and approaches, not absolute truths about the possible location of clandestine graves (Bunch et al. 2017; Cabo et al. 2012; Congram 2010; Congram et al. 2017, 2016; Kolpan and Warren 2017; Molina et al. 2020; Somma et al. 2018).

In recent years, some research has been developed that approach the phenomenon of clandestine graves during the period of extended violence in Mexico. It is important to point out that the works that highlight the use of GIS and remote sensing for the current Mexican case are few and there is still a long way to go on this topic (Guevara et al. 2019; Migues 2019). However, due to the relevance of some cases, such as the Ayotzinapa case in Guerrero, important advances have been made in the use of technology and the construction of geographic models for searches in the country (Silván-Cárdenas 2021; Silván-Cárdenas et al. 2019).

In this work, two main concepts are taken up in order to analyze the problem of clandestine graves in conflict contexts, the first one is based on the work of Congram et al. (2016) in which a new approach to the problem of clandestine graves in conflict contexts is proposed, using spatial clustering tests, "that measures the degree of similarity between things that are clustered in space and correlate of dissimilarity for things that are dispersed" Congram et al. (2016) in addition with spatial clustering some context variables are analyzed. The second concept refers to the model of a "clandestine space", in which areas with a higher probability of finding clandestine graves are analyzed (Silván-Cárdenas et al. 2019).

The rest of the paper is organized as follows. Section 2 describes the study area, its context and data used; Sect. 3 summarizes the methods used for the data analysis, Sect. 4 presents the results from our data analysis and, finally, Sect. 5 emphasizes major conclusions drawn from the study and possibilities for further studies.

2 Description of the Study Area and Context

BC is located in northwestern Mexico, bordering the United States of America (Fig. 1). It has a 71,450 km^2 area and a 3.7 million population (3.0% of the country), 94% of which is concentrated in urban areas and 6% in rural areas, thus contrasting the figures at the national level with 79% and 21%, respectively. The state capital is Mexicali; however, it is doubled in population by Tijuana, with around 2 million people, being part of a much larger metropolitan area that includes the twin city of San Diego, California.

According to Stratfor,[1] the BC area is disputed between the "Sinaloa" cartel conformed by Sinaloa Federation and remnants of Beltran Leyva Organization, Tijuana Cartel and Juarez Cartel and in the other hand "Tierra Caliente" cartel conformed by Cartel Jalisco Nueva Generacion (CJNG) and, remnants of: Knights Templar, la Familia Michoacana and Beltran Leyva Organization. It seems like the remnants of the Beltran Leyva organization are divided, and the CJNG must pay a "piso" for contraband through border crossing, which is under the control of other groups. Furthermore, the results of Stratfor suggest that the larger cartels provide finance and guns to smaller local groups. In this sense, these are some of the reasons why BC and especially the Tijuana Metropolitan Area are the main areas with the highest number of homicides in Mexico.

During 2011 through June 2021, the Tijuana Metropolitan Zone ranks second in homicides, with 8.5% of the national total, just after the Valley of Mexico Metropolitan Zone. In terms of homicide rates per inhabitants, in this same period, the Metropolitan Zone of Tijuana ranks third, after Acapulco (Guerrero) and Tecomán (Colima), with 8.12 homicides per 10,000 inhabitants per year.

From December 2006 to June 2021 a total of 1,098 persons have been reported missing or disappeared in BC, 63% are men, and of the total, just over 50% of the persons correspond to Tijuana. In March 2018, was installed the National Search Commission (Comisión Nacional de Búsqueda-CNB), which is a decentralized administrative body of the Ministry of the Interior (Secretaría de Gobernación), which determines, enforces, and follows up on the search actions for missing and disappeared persons, throughout the national territory, in accordance with the provisions of this Law. Its purpose is to promote the efforts of linkage, operation, management, assessment, and follow-up of the actions between authorities involved in the search, location and identification of persons. From February 2019 to June 2021, the CNB conducted 1,758 search rounds throughout the national territory. BC ranks eighth with 59 search rounds, it is noteworthy that searches have been conducted in the six municipalities of this state (Subsecretaría de Derechos Humanos 2018).

Impunity in BC is also one of the highest in Mexico, according to the Global Impunity Index in Mexico, published by the Center for Impunity and Justice Studies of the University of the Americas Puebla; BC in 2016 and 2018 ranked third worst in the country, with indexes of 74.42 and 78.08, respectively, it is worth mentioning

---

[1] https://worldview.stratfor.com/article/stratfor-mexico-cartel-forecast-2020.

Fig. 1 Location of
clandestine graves in Baja
Caifornia state. *Source*
Elaborated by the authors
with information from Local
Attorney's Office of Baja
California (FGBEBC)

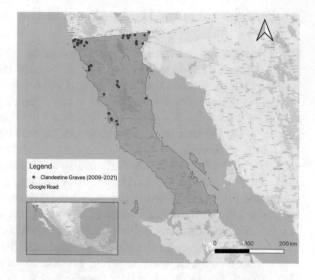

that these figures are above the national average in those years, whose indexes were:
67.42 for 2016 and 69.85 in 2018 (Le Clercq Ortega 2016, 2018).

2.1 Data Used

The data set of hidden graves was acquired through public information access
requests conducted to the Local Attorney's Office of BC (Fiscalía General del Estado
de Baja California-FGEBC).[2] FGEBC provided a spreadsheet of 132 sites where
clandestine graves have been observed from 2009–2021 in the state. From this 132
sites, 52 included latitude and longitude data. These points are shown in Fig. 1.

The travel time and visibility maps where generated following Silván-Cárdenas
et al. (2019), where the urban settlement layer was acquired from the National Institute for Statistics, Geography and Informatics (INEGI), the road networks from the
Mexican Transport Institute (IMT) and open street Maps (OSM), and the digital
elevation model from NASA's shuttle radar topography mission (SRTM).

---

[2] Public information access requests are conducted through National Institute for Transparency,
Access to Information and Protection of Personal Data (INAI-Plataforma Nacional de Transparen-
cia https://www.plataformadetransparencia.org.mx/web/guest/inicio), which is a website from the
Mexican government where citizens can request public information from federal and local authorities, according to the General Law on Transparency and Access to Public Information in Mexico.
Authorities are obliged to respond to the information requests within 20 d. Information can be
granted or reserved, depending on the sensitivity of the information requested or the authority's
interpretation.

3 Methods

3.1 Spatial Point Patterns Analysis

The first step of this study was to conduct spatial point pattern analysis on the grave locations. The objective was to identify spatial patterns that may point to locating new hidden graves based on the distances and clustering of the graves that the FGEBC has previously observed.

Specific tests such as Average Nearest Neighbors (ANN) and Ripley's K function have been suggested by Congram et al. (2016) as appropriate methods to infer these locations in contexts of International or Non-International Armed Conflicts where people have gone missing. We replicated this approach for our study in BC function.[3]

First, ANN was used to check whether there is clustering or dispersion of the 52 georeferenced points of graves in BC, observed during 2009–2021. ANN is a test that produces a Nearest Neighbour Ratio (NRR) based on the average distance from each feature to its nearest neighboring feature. If the ratio is less than 1 there is clustering, whereas a value greater than 1 indicates a dispersed point pattern (Congram et al. 2016).

Second, we implemented the Ripley K function and its L(d) transformation on the 52 points. The test measures the expected number of events within a distance from an arbitrary number of events. It allows the analysis of spatial dependency (clustering) over multiple predefined distance increments or spatial scales (Congram et al. 2016). The test generates random permutations (random distribution of points in an area) to compare against the observed points. We conducted 900 simulations with no correction method to avoid making serious assumptions about the proximity of unknown graves (Congram et al. 2016). We then transformed the K function to the L function to correct the variance increases with the distance.

3.2 Clandestine Space

Congram et al. (2017) have pointed out that there are many spatial variables that could be introduced into a model of grave localization, but few are those that really have to do with the process of creating a grave. In particular, those that explain the clandestine nature of the criminal activity associated with abduction, kidnapping and forced disappearance must be considered. It was under this observation that the concept of Clandestine Space (CS) was recently developed, in which two dimensions are used to characterize it: (1) the spatial accessibility and (2) the spatial privacy (Silván-Cárdenas et al. 2019). These dimensions were, in turn, quantified in terms of travel time and percentage of visibility, respectively. In principle, the CS will

---

[3] ANN was conducted using QGIS and Ripley's K function was conducted using the R programming language.

present a greater probability of containing clandestine graves under the premise that perpetrators choose sites that are quickly accessible and not very visible to the public in order to reduce the risk of being caught.

The CS concept is realized through geospatial modeling, where the geographic space is divided into cells on which both the travel time from an urban settlement and the visibility index are calculated. The first quantity is calculated as an accumulated cost function from the closest urban cell to each point in geographic space, where the cost function is the time it takes to cross the cell. This value is determined from the maximum speeds of the roads, the terrain slope and the fraction of vegetation cover. A slight modification considered in this study was to use only street/road cells with in urban settlement rather than any cell in the urban settlement.

The percentage of visibility of a cell is calculated as the frequency the point is at the line of sight of an observer located on nearby roads, where the range of visibility is limited to a few kilometers (<5 km). The calculation is carried out by accumulating the viewsheds of a digital terrain model with random view points located on the road network. In general, since woody vegetation also inhibits visibility a vegetation layer can also be used. This was not used for BC as this is a semiarid State with very sparse, low vegetation.

Once the dimensions of the CS have been quantified it is possible to use a machine learning method to try inferring the boundary between clandestine and non-clandestine spaces from previously discovered positive and negative burial sites as in Silván-Cárdenas et al. (2019). The alternative pursued in here was to test a deterministic model of the boundary, $B(t, v) = 0$, which is expressed in terms of the travel time (t) and visibility (v) as:

$$B(t, v) = \left(\frac{t}{a}\right)^n + \left(\frac{v}{b}\right)^n - 1 \tag{1}$$

where a is the maximum travel time where a grave have been observed, b is the maximum visibility of a grave site and $n > 0$ is a exponent that controls the curvature and convexity of the boundary. For instance, the boundary is a concave hyperbolic arc for $n = 0.5$, it is a line for $n = 1$, it is a convex circular arc for $n = 2$ and it becomes a rectangle for $n = \infty$. Therefore, a given point in space with time and visibility (t, v) belongs to the CS mask if, and only if, $B(t, v) \leq 0$. In practice, due to outliers present in the data, parameters a and b must be equated to certain percentile, say 99%, value from the data, and the n value can be optimized using some criterion. In this study, we used the number of grave points in the mask divided by the mask total area, which measures the coverage of known grave points per unit area. This measure is hereafter referred to as the grave cover index, or GCI for short.

3.3 Integrating the CS Boundary Model with the Point Pattern Analysis

Although the CS concept allows reducing the search areas, such reduction is generally insufficient to undertake search for clandestine graves, as the this area for an entire State can be as large as 30% of the State area. In this case, results from point pattern analysis can help to further reduce the search area. Specifically, point pattern analysis provides information on the minimum and maximum distances from known grave locations, where the probability of discovering a new grave point is the highest.

With that, we used two circular buffers around each grave location, subtracting the smallest range buffer from the largest one in order to generate a ring for each point. The rings were then merged into a mask which was finally intersected with the CS mask to produce a new integrated search area. An online platform was developed in Google Earth Engine (GEE) Apps to demonstrate the integration of the CS boundary model together with the ring mask analysis (see Fig. 2). This platform allows to interactively adjust the CS boundary model parameters, as well as the minimum and maximum distance of ring mask and provides numeric and graphic information of the mask area and grave points intersected.

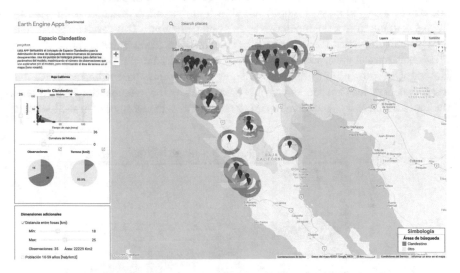

Fig. 2 Online platform that demonstrates the use of the CS boundary model and its integration with other information, including the ring masks from point pattern analysis. *Source* Elaborated by the authors with information from App: https://jsilvan.users.earthengine.app/view/espacio-clandestino

4 Results and Discussion

4.1 Point Pattern Analysis

The ANN test for the BC clandestine graces (0.42 ratio and a Z-Score of −8.009) indicated highly significant clustering, with an observed mean distance of 7 km and an expected mean distance of 16 km. These results suggest that observed clandestine graves in BC are closer to expected if there was a random distribution or dispersion. The L(d) transformation is shown in the Fig. 3, which suggests that new graves in the state of BC may be located at distances between 18 km and 28 km (peak of the L-function) from a known grave point.

4.2 Ring Mask Coverage Analysis

In order to further confirm that the highest probability for discovering a new grave locations is within 18–28 km range away from any given known grave location, we used the GEE App to estimate the GCI, that is the number of points within the mask per unit area of the mask. Figure 4 left shows the variation of the GCI for the ring masks as a function of the mean ring radius and ring width. This result suggests that, on a per unit area basis, there is more chance to discover a new grave within 4 km from previously known grave locations. This seems to contradict the L-function behavior. However, there are a few differences that may explain this apparent discrepancy. First, the L-function does not account for any areal information, whereas the GCI captures the fact that mask area grows much more quickly than the number of grave locations

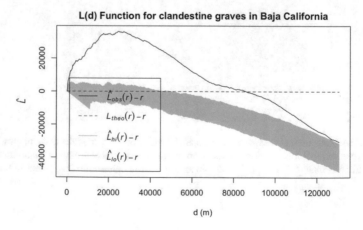

Fig. 3 Expected distance of clandestine graves in BC with L(d) transformation. *Source* Elaborated by the authors with information from Local Attorney's Office of BC (FGBEBC)

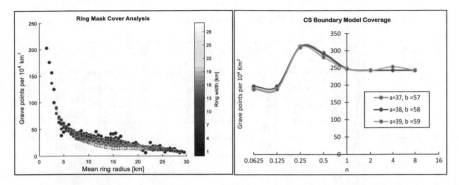

Fig. 4 Variation of the GCI as a function of the mean ring radius (left) and of the CS boundary model parameters (Eq. 1)

as one moves away from a given grave point. Second, the L(d) transformation is based not only on the known grave locations but also on randomly generated points which compensate for bias in the observations. Third, the point pattern analysis uses distances among points regardless of whether they are in distinct metropolitan areas or cities and, as such, it portraits a global pattern. Lastly, the GCI is bi-varied as it used both the minimum and maximum ring radii, whereas the L-function is uni-varied. Furthermore, if the $1/r$ pattern observed for the GCI (Fig. 4 left) is compensated, it results a linear density plot that present three major peaks (data not shown) located around 1–4, 13–18 and 21–26 km, where the latter agrees with the observation from the point pattern analysis.

4.3 Parameters Selection for the CS Boundary Model

As stated earlier, two of the model parameters for the CS boundary model (Eq. 1) can be easily constrained from data, by taking a percentile value of the travel time and visibility of known grave locations. In particular, the BC data presented one outlier with a relatively large visibility (>100), which was disregarded for model parameter estimation. This yielded values for a and b around 38 km and 58, respectively. Then, the exponent parameter n was tested using the CGI criterion. Figure 4 right shows the empirical variation of the GCI as a function of the exponent n of the model. The plots show that the optimal n value is within the range 0.25–0.5 for the BC data. Therefore, all subsequent analyzes used these values.[4]

---

[4] In the GEE App, the exponent is limited to small integer values of $\log_2(n)$.

Table 1 Integrated masks analysis. The $0 - \infty$ range corresponds to no ring masking at all. See the text

CS model parameter	Ring mask range (Km)	Points in mask	Mask area (km^2)	GCI pts/10^4km^2
$n = 0.25$	0-∞	26	840	309.52
	0.1–4	11	72	1527.78
	13–18	21	299	702.34
	18–28	19	461	412.15
$n = 0.5$	0-∞	48	1,708	281.03
	0.1–4	30	153	1,960.78
	13–18	37	648	570.99
	18–28	33	1008	327.38

4.4 Final Search Area

Masks from the CS boundary model with $a = 39, b = 59$ and $n = 0.25, 0.5$ were intersected with three ring masks using the online App. A summary of the analysis of combined masks is presented in Table 1. Overall, it was observed that any of the tested ring masks improved the GCI with respect to using the CS boundary model alone. Furthermore, of the three candidate ranges, the one that seemed more convenient for its high GCI value was the 0.1–4 km range, which reduced the search area to a little less than 10% its initial value.

5 Conclusion

This study have allowed us to further advance the concept of clandestine space that was previously proposed in Silván-Cárdenas et al. (2019), both by providing an explicit model of the boundary and by showing how to integrate information from the point pattern analysis of known grave locations (Congram et al. 2016). Through the integration of geospatial modelling, web mapping, and spatial statistics, we have set forth an online platform that allows to interactively define a search mask that can improve the success of discovering new clandestine grave locations. Results confirmed that information from point pattern analysis can substantially reduce (<10%) the search area provided by the CS boundary model. Being the result from the CS boundary model, the final search areas is warranted to be within practical distances from most urban settlements, 39 min in the case of BC.

Although the point pattern analysis allowed to identify the expected distance where a clandestine grave can be located, it provides no clue about directions thus allowing the search area to grow with the distance. Ultimately, this opens the opportunity for further research to develop and test spatial statistics that explicitly account for the angular variation besides distance, and for an explicit consideration of the variation in point density (or the GCI measure).

References

Bunch AW, Kim M, Brunelli R (2017) Under our nose: the use of GIS technology and case notes to focus search efforts. J Forensic Sci 62(1):92–98

Cabo LL, Dirkmaat DC, Adovasio JM, Rozas VC (2012) Archaeology, mass graves, and resolving commingling issues through spatial analysis. A companion to forensic anthropology, pp 175–196

Congram DR (2010) Spatial analysis and predictive modelling of clandestine graves from rearguard repression of the Spanish Civil War. Ph.D. thesis, Arts & Social Sciences: Department of Archaeology

Congram D, Kenyhercz M, Green AG (2017) Grave mapping in support of the search for missing persons in conflict contexts. Forensic Sci Int 278:260–268

Congram D, Green A, Tuller H (2016) Mapping the missing: a new approach to locating missing persons burial locations in armed conflict contexts. In: Congram D (ed) Missing persons. Multidisciplinary perspectives on the disappeared, pp 207–223. Canadian Scholar's Press Inc., Toronoto, Ontario

Guevara JA et al. (2019) Violencia y terror: hallazgos sobre fosas clandestinas en México 2006–2017

Kolpan KE, Warren M (2017) Utilizing geographic information systems (GIS) to analyze geographic and demographic patterns related to forensic case recovery locations in Florida. Forensic Sci Int 281:67–74

Komar D (2008) Patterns of mortuary practice associated with genocide: implications for archaeological research. Curr Anthropol 49(1):123–133. http://www.jstor.org/stable/10.1086/524761

Le Clercq Ortega JA (2016) Índice global de impunidad en méxico. https://imco.org.mx/indice-global-de-impunidad-de-mexico-2016-via-udlap/. Accessed 01 Sept 2021

Le Clercq Ortega JA (2018) Índice global de impunidad en méxico. https://imco.org.mx/indice-global-impunidad-mexico-2018-via-udlap/. Accessed 01 Sept 2021

Migues DF (2019) Tecnologías de esperanza. apropiaciones tecnopolíticas para la búsqueda de personas desaparecidas en méxico. el caso de las rastreadoras del fuerte. Comunicación y Sociedad, pp 1–29

Molina CM, Wisniewski KD, Drake J, Baena A, Guatame A, Pringle JK (2020) Testing application of geographical information systems, forensic geomorphology and electrical resistivity tomography to investigate clandestine grave sites in Colombia, South America. J Forensic Sci 65(1):266–273

Silván-Cárdenas J (2021) Modelos probabilísticos para el hallazgo de fosas clandestinas, chap. 4. USAID, EnfoqueDH, EAAF, CEDEHM, pp 216–221

Silván-Cárdenas JL, Alegre-Mondragón A, González-Zuccolotto K (2019) Potential distribution of clandestine graves in guerrero using geospatial analysis and modelling. In: Proceedings of the 1st international conferences, vol 13, pp 21–28

Somma R, Cascio M, Silvestro M, Torre E (2018) A GIS-based quantitative approach for the search of clandestine graves, Italy. J Forensic Sci 63(3):882–898

Subsecretaría de Derechos Humanos, P.y.M.: Búsqueda e identificación de personas desaparecidas (2018)

The Geopolitical Repercussions of US Anti-immigrant Rhetoric on Mexican Online Speech About Migration: A Transdisciplinary Approach

Thomas Cattin, Alejandro Molina-Villegas, Julieta Fuentes-Carrera, and Oscar S. Siordia

Abstract This paper presents an ongoing research project that aims to propose a geopolitical analysis of anti-immigrant speech published on the Mexican twitosphere. While Mexico has long defined itself as an emigration country, the apparent growing presence of anti-immigrant discourse online, especially at the Mexican borders, invites us to question the impact of Americans' anti-immigrant speech, bolstered by Donald Trump's election and presidency, on Mexicans' representations. We thus propose a transdisciplinary approach that combines a Convolutional Neuronal Network to detect anti-immigrant speech in geolocalized tweets in Mexican Spanish and a geopolitical diachronic analysis to estimate the relationship between such speeches and Americans' anti-immigrant online representations. With an overall accuracy of 0.76, we are confident that with some improvements the CNN model will be able to detect Mexican anti-immigrant speech on Twitter. We finally discuss that the scope of the analysis would be greatly improved if paired with network and territorial analysis of Mexican anti-immigrant tweets.

Keywords Geopolitical repercussions · United States · Mexico · Twitter · Georeferenced anti-immigrant speech · Hate speech detection

T. Cattin (✉)
Centro de Investigación en Ciencias de Información Geoespacial, Mexico and Institut Français de Géopolitique, Mecico City, France
e-mail: tcattin@centrogeo.edu.mx

A. Molina-Villegas
Consejo Nacional de Ciencia y Tecnología, Centro de Investigación en Ciencias de Información Geoespacial, Mecico City, Mexico
e-mail: amolina@conacyt.mx

J. Fuentes-Carrera
Centro de Investigación en Ciencias de Información Geoespacial, Mecico City, Mexico

O. S. Siordia
Laboratorio Nacional de GeoInteligencia, Centro de Investigación en Ciencias de Información Geoespacial, Mecico City, Mexico

© The Author(s), under exclusive license to Springer Nature Switzerland AG 2022
R. Tapia-McClung et al. (eds.), *Advances in Geospatial Data Science*, Lecture Notes in Geoinformation and Cartography, https://doi.org/10.1007/978-3-030-98096-2_4

1   Introduction

The arrival of several thousands of asylum seekers from Central America at the border town of Tijuana, in November 2018, showed anti-immigration attitudes in a part of the local population. The Mexican anti-immigrant speech was evident online, such as in a Facebook group *Tijuana en contra de la caravana migrante* with up to 4000 members (UnionTJ664 2018), and was instrumentalized by part of the local political elite. The former mayor of Tijuana, Juan Manuel Gastélum Buenrostro, was seen wearing a *Make Tijuana Great Again* cap.

Mexican anti-immigrant speech becoming visible is a problem to be addressed in geopolitical terms: The Mexican territory is the main migratory terrestrial interface for clandestine migrations towards the United States (Casillas 2008) (Fig. 1). Since the September 11th, 2001, terrorist attacks, the Mexican borders have become geostrategic externalization territories for the United States border controls. This was done by joint policies alongside the Mexican government, such as the Southern Border Program (*Programa Frontera sur*) in 2014. All while the clandestine migration notably from Central America increased (Capps et al. 2019), these policies have transformed the Mexican borders into migratory buffer zones where large groups of migrants get trapped for several months waiting for the resolution of their migratory processes.

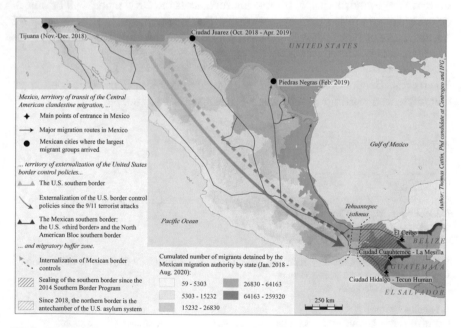

Fig. 1 Mexico: migration interface and migratory buffer zone of the United States (Casillas 2008; UPM 2020; Zepeda and Fuentes Carrera 2020)

Mexico does not yet seem to have the actors and mechanisms favoring the emergence of intense and lasting anti-immigrant speech, such as the political instrumentalization of the subject by opposed partisan blocs (Ernst et al. September 2017). However, expressions such as *"Make Tijuana Great Again"* or the hashtag *#Mexico-Primero*, derived from Donald Trump's campaign slogans which aimed to electorally exploit inter-ethnic power rivalry in the US, indicate some interiorization of American anti-immigrant rhetoric in Mexico. Such interiorization invites us to think about Mexican anti-immigrant speech not as a cultural but as a geopolitical phenomenon where the migratory situation in Mexico, a byproduct of US migration policies, and the electoral power struggle in the US have bolstered the online transnationalization of American anti-immigrant speech.

There are two objectives to this paper. First, to detail the approach used to build a Convolutional Neuronal Network (CNN) model able to detect anti-immigrant speech in geolocalized Mexican Spanish tweets and to present the preliminary result of said model. Second, to propose a transdisciplinary methodology, based on the data obtained from the classification, to verify the hypothesis that American anti-immigrant speech online during Donald Trump's presidency has had a quantitative and qualitative impact on Mexican anti-immigrant tweets.

2   State of the Art

2.1   State of the Art of the Online Analysis of Anti-immigrant Speech

With the emergence of the World Wide Web, social media, and in particular, Twitter, have imposed themselves as privileged media platforms for verbal opposition, especially on migration issues, between users that interact almost instantaneously through the content they produce (Karatzogianni et al. 2016; Koylu et al. 2019). An essential takeaway from these studies is the transnational aspect of the debate on social media (Ferra and Nguyen 2017; Nguyen 2016). Toudert recently analyzed the conflictual nature of the debate on Twitter about the arrival in Tijuana of the Central American migrant caravan (Toudert 2021). In most works, the sentiment analysis is limited to surface forms, such as lexicon and syntax. We argue that developing a classification model based on deep learning techniques can improve the detection of anti-immigrant speech. Furthermore, the use of geolocalized data from Twitter will enable us to isolate and analyze specifically the Mexican anti-immigrant speech.

2.2   State of the Art of Online Hate Speech Detection

The recurrent presence of hate speech on social media has generated an interest in the automatic detection of this type of content by using Natural Language Processing (NLP) techniques (Fortuna and Nunes September 2018). Recent works have demonstrated the efficiency of an approach based on neuronal networks to detect hate speech in English (Ridenhour et al. 2020; Zhang et al. 2018) and Spanish (Pereira-Kohatsu et al. 2019). Other research has focused on certain types of hate speech such as misogyny (Molina-Villegas 2021). Even if the detection of anti-immigrant hate speech in Spanish is of undeniable interest (Basile et al. 2019), this domain is still unexplored. This project aims at building a model able to detect anti-immigrant speech in Spanish in the Mexican territorial context.

3   Anti-immigrant Speech Detection Using CNN

3.1   Twitter Georeferenced Data and How to Obtain It

Twitter data presents the considerable advantage of being relatively easy to access with the platforms' Application Programming Interfaces. The basic data unity is the tweet, a short message (280 characters) accompanied by metadata such as username, date, language, geolocalization, and information about the interaction of other users (retweets, comments, mentions, and favorites). In our case, the detection of anti-immigrant hate speech comes from Machine Learning models specifically created to analyze patterns in the textual information of the previously collected tweets based on geolocalization criteria.

A big data set of tweets published in Mexico between January 2017 and May 2021 were collected using the *Autómata Geointeligente en Internet* (AGEI) developed by Centrogeo[1] and previously used for another research work (López-Ramírez et al. 2019). Due to the fact that only about 1 to 2 percent of all tweets are georeferenced, and about 1 percent of all tweets contain hate speech (Pereira-Kohatsu et al. 2019), the percentage for anti-immigrant speech is probably much lower, it is necessary to dispose of a large raw database.

Once downloaded, the AGEI data was mapped in a Geographical Information System (GIS) and regrouped by year, three-month periods, and by region of publication (Fig. 2). These regions correspond to northern Mexican border states, the southern border states, the rest of the Mexican states, and the US southern border region. To remediate possible imprecisions in tweets' geolocalisation, we applied a buffer of 20km to the four regions.

---

[1] https://www.centrogeo.org.mx/geointeligencia.

3.2 Manual Tweet Labeling

The first step to build a supervised classification model is the most time-consuming one: labeling a large number of tweets manually into our two classes that are anti-immigrant and those that are not (Fig. 2). However, to speed up this process, the tweets collected with AGEI were filtered using a catalog of 49 words and expressions susceptible to anti-immigrant speech. In parallel, a Twitter search by context and users was programmed, a tweet published in a specific context or by a specific user where interactions with the tweet could contain anti-immigrant speech.

The tweets were then manually labeled according to five criteria. These criteria were elaborated from the geopolitical conceptualization of anti-immigrant speech. The criteria were sufficiently simple to allow non-specialists to use them and general enough to detect anti-immigrant speech published on Twitter in Mexico. The criteria were qualitatively validated by a panel of experts. To further improve the objectivity of criteria, it is planned to measure statistically the inter-rater agreement based upon experts' individual classification, using the criteria, of a small data set containing both anti-immigrant and not anti-immigrant tweets.

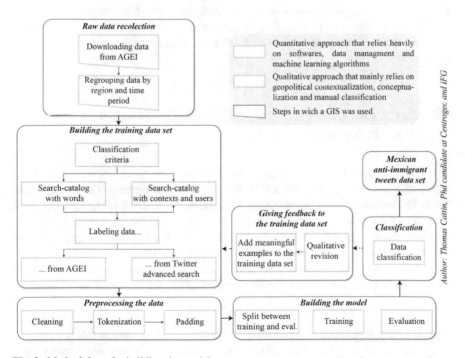

Fig. 2 Methodology for building the model

3.3   Model Training and Classification

Before training the model and classifying the data, they were submitted to a series of transformations to generate representative features, which served as inputs in the model (Fig. 2). These operations were done in the following order: data cleaning, tokenization, and padding. Labeled and pretreated data were then divided into two data sets to train and evaluate the model. Because of the imbalance between negative (4548) and positive (1073) tweets in our training data, 18.5 percent of the positive tweets (200), and the same number of negative tweets, were used for the evaluation in order to measure how accurate the model is for both classes.

The training data set was then used to train a Convolutional Neuronal Network (Fig. 3). This is a technique primarily used in image processing, but applied to texts, a CNN is very effective in applying successive filters that reduce the entry vectors' dimensions (embeddings) while increasing the effectiveness of finding text patterns (parameters). This type of model has proven efficient in hate speech detection (Molina-Villegas 2021; Zhang et al. 2018) and Geographic Named Entity Recognition (Molina-Villegas et al. 2021).

Once trained, the model was used to classify pretreated data collected from AGEI. The earliest classification was used as feedback for the training stage (Fig. 2). The qualitative revision of the classified data proved useful for adding negatives and positives to the training data set by manually reclassifying false positives and false negatives.

3.4   Results on Classification Model not Anti-immigrant, Anti-immigrant

The Table 1 presents the preliminary results for the statistical performance of the binary classification model (not anti-immigrant, anti-immigrant). The average precision for the not anti-immigrant class indicates that we still have a relatively high number of false positives, which was corroborated during the qualitative revision of the classification output. Whereas the recall for the anti-immigrant class indicates that the model fails to detect a good portion of the anti-immigrant tweets.

To further improve the model, we need to increase the training data set with both anti-immigrant and not anti-immigrant tweets. Our current training data set contains

Table 1 Results of binary classification for anti-immigrant speech in Mexican tweets

	Accuracy	Precision	Recall	F-measure	support
Not anti-immigrant		0.7219	0.8700	0.7891	200
Anti-immigrant		0.8364	0.6650	0.7409	200
All	0.7675				400

only 1073 anti-immigrant tweets and mainly consists of data from the northern and southern border regions in 2018 and 2019. It is therefore neither geographically nor temporally representative of the entire data set. This could explain the high number of false positives found during a qualitative revision of newly classified data. It will also be necessary to experiment with the variation of specific parameters of the model, such as the number of filters (300), the number of neurons in the hidden layer (300), and the dropout rate (0.45).

4   A Geopolitical Analysis of Mexican Anti-immigrant Speech on Twitter During the Trump Administration

It now has to be said that information obtained from Twitter does not represent the whole anti-immigrant speech in Mexico. Those who publish on Twitter have internet access for which Mexico has a marked urban/rural fracture (INEGI 2020), and they are relatively politicized as Twitter appears to be the preferred platform for interactions between the ruling class and the governed (Bruns and Stieglitz 2014; Tandoc andJohnson 2016). Furthermore, georeferenced tweets represent a tiny fraction of all tweets, and those who publish geographical data are not representative of the wider Twitter population (Sloan and Morgan 2015). In this section, we do not pretend to extend the analyses outside of Twitter for those reasons.

In recent years, Donald Trump's ability to steer the terms of the migration debate on Twitter has been demonstrated (Koylu et al. 2019). Through his @realDonaldTrump Twitter account, he instrumentalized the core elements of American racist discourse now aimed at those who come from beyond the southern US border (Haney-López 2014). We thus propose to use @realDonaldTrump's anti-immigrant tweets to estimate by proxy the influence of US anti-immigrant discourse on Mexican anti-immigrant speech (Fig. 3).

In geopolitics, the diachronic analysis revolves around the division of time into crises and ruptures that highlight the evolution of a phenomenon on the territory. We can transpose this analysis method on Twitter by dividing the publication history of @realDonaldTrump into temporal milestones, tweets that crystallize the evolution of Donald Trump's discourse on migration. The @realDonaldTrump tweets published between January 2017 and January 2021 were downloaded from the Trump Twitter Archive[2] and the ones related to migration were manually selected. Based on personal knowledge, we then determine the temporal milestones for Trump's domestic and international migration tweets.

Once validated, those temporal milestones will be used as an analysis grid for Mexican anti-immigrant tweets obtained from the classification (Fig. 3). We established two ways to infer repercussions of American anti-immigrant speech on those tweets.

---

[2] https://www.thetrumparchive.com.

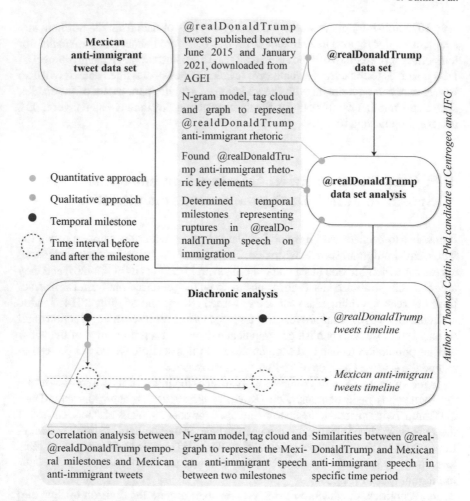

Fig. 3 Roadmap for diachronic analysis

Firstly, by examining over the period the correlation between the @realDonaldTrump milestones and the variations in the number of Mexican anti-immigrant tweets. Secondly, by looking for lexical similarities between Mexican speech and @realDonaldTrump's using NLP techniques like n-gram models. With those two analyses, we should have a somewhat accurate picture of how American anti-immigrant rhetoric, Donald Trump's or similar rhetoric used by users unknown to us for the moment, impacted both quantitatively and qualitatively the Mexican anti-immigrant speech on Twitter.

5 Conclusions

When completed, this research project will make several contributions to both NLP and geopolitics communities. In addition to providing a concrete application of NLP techniques in social sciences, we will have designed a data base and a model for detecting anti-immigrant speech in Spanish. Regarding geopolitics, this work will provide evidence that, through digital social networks, highly politicized and sensitive debates in one territory can be reshaped by the repercussions of similar debates in entirely different territorial contexts. Finally, we hope that this project will serve as an example of how useful a transdisciplinary approach can be when the exponential growth of available data and the emergence of the internet as a real digital space offer an unprecedented opportunity to deepen our knowledge about social and territorial interactions.

There is still a lot to be done and we argue that this research would be greatly improved if combined with other analyses. Network analyses based on graph theory techniques have proved successful in studying interactions between digital social network users and their consequences (González-Bailón 2017), including in geopolitics (Gérard and Marotte 2020). Based on interactions such as mentions, replies, or favorites, we could reconstruct the network that underlies the Mexican anti-immigrant speech on Twitter. With a topographic analysis of this network, we could identify the main users and communities of users who disseminate those speeches and how they bridge the boundaries of the Mexican and American national debate about migration.

One of the major contemporary geopolitical questions is to understand how interactions within digital spaces impact the production of territories and *vice versa*. To investigate this relationship, we could use the network analysis as an input for two field studies, one in Tijuana (northern Mexico border) and the other in Tapachula (southern Mexico border). Where possible, local actors of those two border cities whose digital *alter ego* plays a role in the Mexican anti-immigrant community on Twitter, and whom we manage to identify, would be questioned about their views on immigration and their presence online during qualitative interviews.

References

Basile V, Bosco C, Fersini E, Debora N, Patti V, Pardo FM, Rosso P, Sanguinetti M (2019) SemEval-2019 task 5: multilingual detection of hate speech against immigrants and women in twitter. In: Proceedings of the 13th international workshop on semantic evaluation, pp 54–63, Minneapolis, Minnesota, USA, 2019. Association for Computational Linguistics

Bruns A, Stieglitz S (2014) Twitter data: what do they represent? Inf Technol 56(5):240–245

Capps R, Meissner D, Ruiz S, Ariel G, Bolter J, Pierce S (2019) From control to crisis: changing trendrs and policies reshaping U.S.-Mexcio border enforcment. Technical report, Migration Policy Institute, Washington, DC, 2019

Casillas R (2008) Las rutas de los centroamericanos por México, un ejercicio de caracterización, actores principales y complejidades. Migración y desarrollo 10:157–174

Ernst N, Engesser S, Büchel F, Blassnig S, Esser F (2017) Extreme parties and populism: an analysis of Facebook and Twitter across six countries. Inf Commun Soc 20(9):1347–1364

Ferra I, Nguyen D (2017) #migrantcrisis: & "tagging" the European migration crisis on twitter. J Commun Manag 21(4):411–426

Fortuna P, Nunes S (2018) A survey on automatic detection of hate speech in text. ACM Comput Surv 51(4):1–30

González-Bailón S (2017) Decoding the social world: data science and the unintended consequences of communication. Information policy series. MIT Press, Cambridge, MA

Gérard C, Marotte G (2020) #AffaireBenalla : déconstruction d'une polémique sur le rôle de la communauté Twitter ≪ russophile ≫ dans le débat politique français. Hérodote, N o 177–178(2):125

Haney-López I (2014) Dog whistle politics: how coded racial appeals have reinvented racism and wrecked the middle class. Oxford University Press, Oxford. OCLC: 884873319

INEGI (2020) En méxico hay 84.1 millones de usuarios de internet y 88.2 millones de usuarios de teléfonos celulares: Endutih, (2020) Technical report. Mexico, Instituto Nacional de Estadística y Geografía, Mexico City, p 2021

Karatzogianni A, Nguyen D, Serafinelli E (eds) (2016) The digital transformation of the public sphere: conflict, migration, crisis and culture in digital networks. Palgrave Macmillan UK

Koylu C, Larson R, Dietrich BJ, Lee K-P (2019) Carsentogram: geovisual text analytics for exploring spatiotemporal variation in public discourse on twitter. Cartogr Geogr Inf Sci 46(1):57–71

López-Ramírez P, Molina-Villegas A, Siordia OS (2019) Geographical aggregation of microblog posts for LDA topic modeling. J Intell Fuzzy Syst 36(5):4901–4908

Molina-Villegas A (2021) La incidencia de las voces misóginas sobre el espacio digital en méxico. In: Pérez-Barajas AE, Arellano-Ceballos AC (eds) Jóvenes, plataformas digitales y lenguajes: diversidad lingüística, discursos e identidades. Elementum, Mexico (in press)

Molina-Villegas A, Muñz-Sanchez V, Arreola-Trapala J, Alcóntara F (2021) Geographic named entity recognition and disambiguation in mexican news using word embeddings. Expert Syst Appl 176:114855

Nguyen D (2016) Analysing transnational web spheres: the European example during the Eurozone Crisis. Palgrave Macmillan UK, London, pp 211–233

Pereira-Kohatsu JC, Quijano-Sónchez L, Liberatore F, Camacho-Collados M (2019) Detecting and monitoring hate speech in twitter. Sensors 19(21):4654

Ridenhour M, Bagavathi A, Raisi E, Krishnan S (2020) Detecting online hate speech: approaches using weak supervision and network embedding models. In: Thomson R, Bisgin H, Dancy C, Hyder A, Hussain M (eds) Social, cultural, and behavioral modeling. Series title: lecture notes in computer science, vol 12268. Springer International Publishing, Cham, pp 202–212

Sloan L, Morgan J (2015) Who tweets with their location? Understanding the relationship between demographic characteristics and the use of geoservices and geotagging on Twitter. PLoS ONE 10(11)

Tandoc EC, Johnson E (2016) Most students get breaking news first from Twitter. Newspaper Res J 37(2):153–166; Publisher: SAGE Publications Inc

Toudert D (2021) Crisis de la caravana de migrantes: algunas realidades sobre el discurso público en Twitter. Migr Int 12

UnionTJ664 (2018) Tijuana en contra de la caravana migrante

UPM (2020) Eventos de extranjeros presentados ante la autoridad migratoria, según entidad federativa y municipio (2020) Technical report. Mexico, UPM-Secretaría de Gobernación, Mexico City, p 2021

Zepeda B, Fuentes-Carrera J (2020) La frontera México-Guatemala y el perímetro de seguridad de Estados-Unidos 2000-2020. In: Fuentes Carrera J (ed) Entre lo político y lo espacial: representaciones geopolíticas de la región transfronteriza México-Guatemala. Región Transfronteriza México Guatemala, pp 49–84

Zhang Z, Robinson D, Tepper J (2018) Detecting hate speech on twitter using a convolution-GRU based deep neural network. In: Gangemi A, Navigli R, Vidal M-E, Hitzler P, Troncy R, Hollink L, Tordai A, Alam M (eds) The semantic web. Springer International Publishing, Cham, pp 745–760

Spatial Analysis of a Forest Socio-Ecological System in Oaxaca, Mexico Based on the DPSIR Framework

José García-Hernández and Iskar Jasmani Waluyo-Moreno

Abstract The Driver-Pressure-State-Impact-Response (DPSIR) framework can be used to analyze interactions between society and the environment, including for forest socio-ecological systems, such as ones in Oaxaca, Mexico. In this paper, the DPSIR framework was operationalized through indicators obtained from Inegi, Semarnat and Conafor: the economic value of agricultural activities and population were considered Drivers (D), surfaces used for agricultural activities were considered Pressure (P), current land use and vegetation coverage was used to describe State (S), timber and non-timber production was considered Impacts (I) and payments for ecosystem services (PES) were considered Response (R). Values per municipality of each of the indicators allowed estimating the spatial autocorrelation (Moran's I) between indicator groups. Results suggest that municipalities with agricultural and livestock activity are mostly located in municipalities that border Veracruz and in certain parts of the Costa. Municipalities with more timber production are located in the Sierra Norte (Ixtlan de Juarez, Santa Catarina Ixtepeji, San Pablo Macuiltianguis). In Oaxaca, agricultural and livestock production cluster in similar regions, thus little trade-offs. Forestry shows a mixture of clusters which may indicate forest fragmentation. Results are indicative of the effects of PES that promote forest conservation in Sierra Norte and patterns of agriculture and livestock near borders which may be indicative of trade activities.

Keywords Deforestation · DPSIR framework · Spatial analysis

J. García-Hernández (✉)
Universidad de Chalcatongo, Avenida Universidad S/N Col. Centro, Chalcatongo de Hidalgo, Tlaxiaco, CP. 71100, Oaxaca, Mexico
e-mail: jgarcia@unicha.edu.mx

I. J. Waluyo-Moreno
Centro de Investigación en Ciencias de Información Geoespacial A.C., Contoy 137, Col. Lomas de Padierna, Alcaldía Tlalpan, CP. 14240, Cdmx, Mexico, Mexico
e-mail: iwaluyo@centrogeo.edu.mx

© The Author(s), under exclusive license to Springer Nature Switzerland AG 2022
R. Tapia-McClung et al. (eds.), *Advances in Geospatial Data Science*, Lecture Notes in Geoinformation and Cartography, https://doi.org/10.1007/978-3-030-98096-2_5

1 Introduction

From 1980 to 2001 Oaxaca lost 511,361 hectares (4%) of its forest, more than 24,000 hectares per year (Velázquez et al. 2003). This trend continued from 2001 to 2012 as an additional 232,157 hectares (3.6%) were lost at a rate of 16,583 hectares per year (Hansen et al. 2013). Granted the rate of deforestation has reduced, it continues to be an challenge for sustainable development. Some of the main reasons for deforestation are agriculture, livestock, fires, urbanization and communication infrastructure development (Ellis et al. 2016). Although there have been efforts to mitigate these patterns through public policies, in Mexico many of these policies have contradicted themselves by simultaneously promoting conservation and agricultural development. The relationship between agricultural, livestock production and deforestation are generally accepted as true. Since our current food system is dependant of these activities, we are faced with finding a balance between human development and forest conservation which is, at the very least, complex. In this regard, we consider it important to develop easy to understand visual tools that help identify where, when and why these relationships seem to be more prevalent.

1.1 Oaxaca's Forest Socio-Ecological System

Oaxaca Mexico covers 93,241 square kilometers which is politically divided into 570 municipalities and 1632 *ejidos* (communal agricultural communities). Close to 90% of forests in Oaxaca are communal properties (Bray et al. 2005), managed for timber production, community entrepeneurship and conservation (Bray et al. 2008, 2005; Martin et al. 2011). According to the State Forestry and Land Use Inventory (Inventario Estatal Forestal y de Suelos) (Inventarios forestales 2011), 67.3% of the state is covered by forests (6,295,473.77 ha) and 32.9% (3,100,503.96 ha) are farming, human settlements, urban zone, bodies of water and areas with no vegetation. Furthermore, it is considered one the most biodiverse states which is home to approximately 845 vascular plants, 190 mammals, 736 birds, 245 reptile and 1103 butterfly species (García Mendoza et al. 2014). Oaxaca is also culturally diverse with 16 ethnic groups, many of which suffer from high levels of social marginalization. It is a region with intense agricultural and livestock production, as well as, of high interest for mining companies.

This combination of environmental assets, prehispanic cultures, communal land management and economic interests make Oaxaca's forest socio ecological system an important cultural and environmental asset that is difficult to evaluate. One possible approach is the DPSIR framework because Oaxaca is a clear example of a Socio-Ecological System (SES) where natural ecosystems are linked to and affected by one or more social systems which encompasses interactions between cultural, political, social, economic, ecological, technological aspects and others (Anderies et al. 2004). According to Ostrom and in broad terms an SES consists of a society who interact

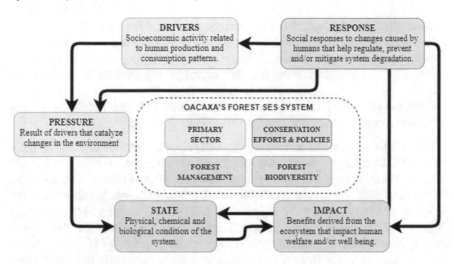

Fig. 1 Combination Forest SES and DPSIR framework

with ecosystems and extract from resource systems. Society also works to maintain and conserve the resource systems with rules and procedures that are parts of a Governance System related to Ecological Systems (ECO) in a Social, Economic and Political Setting (S) (Ostrom 2009). In this study, we consider four components of Oaxaca's forest SES (see Fig. 1).

1. **Primary sector**—Productive activities related to forestry, agriculture and ranching. This is also related to population.
2. **Conservation efforts and policies**—In the form of Payments for Ecosystem Services (PES)
3. **Forest management**—Benefits derrived from forestry in the region.
4. **Forest biodiversity**—Primary (forests, jungles and mangroves) and secondary vegetation (shrubland and savannah).

1.2 The DPSIR Framework

The DPSIR framework is often classified as a policy framework that exposes the interaction between social and ecological systems (See internal part of Fig. 1) (Binder et al. 2013). It breaks down socio-ecological systems into (Burkhard and Müller 2008):

- **Drivers (D)**—Socioeconomic activity that correspond to human production and consumption patterns.
- **Pressure (P)**—Pressures are the result of Drivers and catalyze changes in the environment such as land use changes and natural resource extraction.

- **State (S)**—The physical, chemical and biological condition of an SES.
- **Impact (I)**—Benefits derived from ecosystem services which can have an impact on human welfare or well-being. These include provisioning of resources, regulation of natural habitats cultural benefits and supporting services that maintain an SES.
- **Response (R)**—Society responds to changes caused by human activity in order to regulate, prevent and/or mitigate (or not) SES degradation.

The DPSIR framework can be used to simplify complex systems, such as Oaxaca's forest SES, by breaking them down into the relationships between its components which minimizes communication gaps between stakeholders (Hanne Svarstad et al. 2008). The operationalization of this framework may help expose possible courses of action to attend environmental problems (Cumming 2014) and cause-effect relationships between social, economic and environmental components of a system allowing the visualization of potential issues between them (Pinto et al. 2013). It has been utilized for qualitative forest research in Austria to identify the best actions for forest management (Vacik et al. 2007), fynbos pest and disease control in South Africa (Roura-Pascual et al. 2009), forest restoration in the Grijalva basin in Mexico (Ramírez-Marcial et al. 2014, dry ecosystem analysis in Latin America and for social responses to deforestation and soil degradation between 1990–2010 (Newton et al. 2012). In China, it was used to analyze the influence of socioeconomic activities on biodiversity, ecosystem services and human wellbeing, (Hou et al. 2014); to evaluate water poverty (Sun et al. 2018); to know the principal factors involved in CO^2 emissions at national and regional scale (Wei et al. 2019) and to analyze security and resilience (Nathwani et al. 2019). Finally in Spain, it was used to evaluate water management policy (Vidal-Abarca et al. 2014) and conservation policy (Santos-Martín et al. 2013).

In this research we propose using a combination of these models as seen in Fig. 1.

1.3 Mapping DPSIR Indicators

Interactive web maps may be a cost effective way to facilitate decision-making that allow quick and simple information access for policymakers. Hosting data in free open-source platforms such as Github and deploying maps using R software available allows developing low cost web maps for specific reasons (Moreno et al. 2021). Software libraries for R have been created for spatial analysis, making it a powerful open source tool capable of reading, manipulating and mapping georeferenced information (Mas 2018). The proposed model (see Fig. 1) combined with spatial analysis may provide the basis for a systematic evaluation and visualization of complex systems, such as Oaxaca's forest SES. Exploratory analysis of geospatial data based on the DPSIR framework combined with spatial autocorrelation has been used to identify provinces with hydric poverty poverty in China (Sun et al. 2018), with kenel density estimation model (Liu et al. 2019) and the influence degree of different factors on CO_2 emission in different regions of China (Wei et al. 2019).

In this sense, we used accessible data to develop an open access geographical information platform which presents information regarding Oaxaca's forest SES. This an effective and low cost decision making tool for regions with few resources for such tools and where there is a constant struggle between conservation and agriculture such as Oaxaca. Our objective is to perform an exploratory analysis of thematic maps and the spatial autocorrelation of DPSIR indicators in order to quantify them and identify spatial clusters in regions and municipalities. The selection of indicators was based on a review of literature and are explained in Sect. 2.

2 Materials and Methods

2.1 Data and Software

Data was acquired from three different sources (1) State and Municipal Database (SIMBAD) of the National Institute of Statistics and Geography (INEGI)[1], (2) National Envrionmental and Natural Resources Information System (SNIARN) and (3) National Forest Commision (CONAFOR)[2]. It was prepared using QGIS version 3.20.2 and R version 4.1.1. Geospatial data was manipulated using GEODA software and ddply libraries for R and visualizations were created using Leaflet and Shiny libraries.

2.2 Procedure

Data acquired in 2016 was submitted to a three stage procedure: 1) exploratory and confirmatory factorial data analysis, 2) DPSIR model design and 3) Indicator and spatial autocorrelation mapping.

- **Stage 1: Exploratory and confirmatory factorial analysis**—A total of 31 indicators were proposed based on previous studies (Hou et al. 2014; Ramírez-Marcial et al. 2014; Roura-Pascual et al. 2009; Santos-Martín et al. 2013; Tsai et al. 2009; Vacik et al. 2007; Vidal-Abarca et al. 2014), only 11 indicators were selected considering statistical significance and data availability.

---

[1] Sistema Estatal y Municipal de Bases de Datos (SIMBAD) del Instituto Nacional de Estadística y Geografía (INEGI)

[2] Sistema Nacional de Información Ambiental y de Recursos Naturales (SNIARN), Comisión Nacional Forestal (CONAFOR)

Table 1 DPSIR Indicators

DPSIR	Indicators	Description
Driver (D)	Production values of timber forest production, non-timber forest production, agricultural production, livestock production and population	The higher the values, the more attractive the activity is in the region
Pressure (P)	Percentage of land with active agricultural activity	The higher the percentage of total surface area occupied with agricultural activities, the higher the pressure on the ecosystem
State (S)	Percentage of land with primary vegetation (forests, jungles and mangroves), Percentage of land with secondary vegetation (mostly shrubland and savannah)	The higher the percentage of non-human induced vegetation is considered better
Impact (I)	Production volume of timber forest production, non-timber forest production	The higher production from forest activities, the more benefits are obtained directly from forests
Response (R)	Area that receive Payments for Ecosystem Services (PES)	The more the area receives PES, the more likely its non-human induced vegetation will be conserved

- **Stage 2: DPSIR model design**—The DPSIR framework was operationalized through indicators that remained after the exploratory and confirmatory data analysis were grouped according to the DPSIR framework as shown in Table 1.
- **Stage 3: Indicator and spatial autocorrelation mapping**—Values for each of the previous indicators per municipality were used to create thematic maps of each of the indicators and their corresponding spatial autocorrelations tests performed with GEODA. Results obtained were then mapped using R with a number of libraries for data manipulation and spatial polygon visualization.

Exploratory spatial data analysis (ESDA) is used to explore the spatial dependence and heterogeneity of data (Anselin 1995), and normaly consists of two parts: global and local spatial autocorrelation. Global spatial autocorrelation describes the spatial attribution of objects by disclosing the similarity between spatially adjacent phenomena through the global Moran Index which ranges from -1 to 1 (Table 2). Positive/negative values of Moran's I indicate a positive/negative spatial autocorrelation between objects, and a value of zero indicates a lack of spatial autocorrelation. Local Indicators of Spatial Association (LISA) express whether or not elements of a certain attribute value are adjacent to each other. These are often summarised by classifying them from 0 to 4 as explained in Table 2.

Table 2 LISA clusters

Value	Description
No clustering (0) (NC)	Random spatial distribution of a given indicator. No adjacency
High–High (H–H) clustering (1)	Adjacency that results from areas with HIGH levels of a given indicator surrounded by areas with HIGH levels of the same indicator
Low–Low (L–L) clustering (2)	Adjacency that results from areas with LOW levels of a given indicator surrounded by areas with LOW levels of the same indicator
Low-High (L-H) clustering (3)	Adjacency that results from areas with LOW levels of a given indicator surrounded by areas with HIGH levels of the same indicator
High-Low (H-L) clustering (4)	Adjacency that results from areas with HIGH levels of a given indicator surrounded by areas with LOW levels of the same indicator

3   Results

Maps of all indicators and their corresponding autocorrelation were created and available online. The following images are screenshots of the online interactive maps that can be deployed locally by running the code available in Github and/or through Shiny Apps.[3]

3.1   Overview of DPSIR Thematic Maps

In 2016, the highest timber forest production values were concentrated in the municipalities of Ixtlan de Juarez in the Sierra Norte, Santa Cruz Itundujia in the Mixteca and Santiago Textitlan in the Sierra Sur of Oaxaca, and high non-timber forest production were found in the municipalities of San Miguel Chimalapa on the Istmo of Tehuantepec and in Santiago Pinotepa Nacional on the Costa of Oaxaca. High crop production values (13 of 570 municipalities) in the Papaloapan basin (Acatlan de Perez Figueroa, San Miguel Soyaltepec, San Juan Bautista Tuxtepec, Loma Bonita, Santiago Yaveo) Sierra Norte (San Juan Cotzocon, San Juan Mazatlan), Istmo of Tehuantepec (Matias Romero Avendaño and San Pedro Tapanatepec), and finaly on the Costa (Villa de Tututepec, Santiago Jamiltepec, Santa Maria Huazolotitlan, Santiago Pinotepa Nacional). Matias Romero Avendaño is the municipality with the highest livestock production value. The most populated areas are the capital and San Juan Bautista Tuxtepec. Muncipalities with most agricultural land cover where San Juan Bautista Tuxtepec, Cosolapa, and San Pedro Ixcatlan. Municipalities with most primary vegetation are in the Sierra Sur and Costa, and high secondary vegetation

---

[3] https://github.com/iskarwaluyo/dpsir_autocorrelation_oaxaca_forest.

was concentrated in Valles Centrales and Mixteca. Municipalities with the most timber forest production are in the Sierra Sur, Sierra Norte and Mixteca. Municipalities
with most payments for ecosystems services are in the Sierra Norte and Cañada (see
Fig. 2).

3.2   Overview of DPSIR Spatial Autocorrelation

There is a high concentration of agricultural drivers in the municipalities of the
Papaloapan region which are valleys with pineapple producers (Moran's I = 0.036).
Meanwhile, there are low–low clusters in the municipalities of Mixteca, Sierra Sur
and Istmo of Tehuantepec. Livestock activities (Moran's I = 0.356) clustered on the
border with the state of Veracruz, municipalities that belong to the regions of the Istmo
of Tehuantepec, and Papaloapan and in some municipalities on the Costa. Population
high–high clusters with a Moran's I of (Moran's I = 0.203) are grouped in the
municipalities of the Papaloapan region, Valles Centrales and in some municipalities
of the Istmo of Tehuantepec.

   On the other hand, timber does not seem to cluster statewide (Moran's I = 0.020).
However, there are some high–high clusters located in the northern and southern
sierra and Istmo of Tehuantepec. There were no low–low clusters. Non-timber production displayed a slightly negative spatial autocorrelation statewide (Moran's I =
−0.006) . There are no high–high clusters and the low–low cases are scarce and
scattered.

   The percentage of land used for agriculture, considered Pressure (P), shows positive global spatial autocorrelation (Moran's I = 0.467). Many low–low clusters are
located along a central strip that passes through the state from its northern border to
its southern coast, and a small high–high cluster on the northern tip of the state.

   Primary vegetation shows positive spatial autocorrelation (Moran's I = 0.137).
High—high clusters are in Sierra Sur and low–low in some municipalities of Valles
Centrales and the Costa (Pinotepa Nacional). Secondary vegetation displayed posi-
tive spatial autocorrelation (Moran's I = 0.270) in the Mixteca, Sierra Sur and Valles
Centrales, however, most are low–low clusters.

   Regarding Impact, timber production shows some degree of autocorrelation
(Moran's I = 0.188); high–high clusters are located in the Sierra Norte and Sierra
Sur of Oaxaca and the low–low clusters in some municipalities of the Mixteca and
Istmo of Tehuantepec. Non-timber forest production (Moran's I = 0.041) does not
present spatial autocorrelation. The high–high clusters in municipalities of Mixteca
and Papaloapan surrounded by low-high clusters.

   Response, evaluated through the percentage area that generated PES show a high
spatial autocorrelation (Moran's I = 0.656). High–high clusters are mostly found
in the Sierra Norte of Oaxaca and low–low ones in the Istmo of Tehuantepec, Mixteca, part of the Valles Centrales and the Costa. The high concentration of PES are
located in areas with less clusters of agricultural and livestock activity (see Fig. 3 and
Table 3).

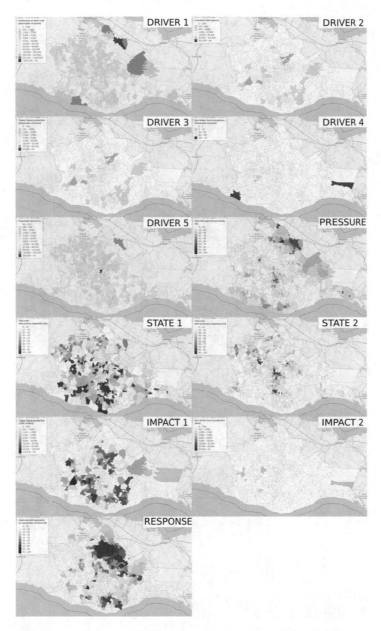

Fig. 2 Indicator maps: Driver 1—Timber forest production value, Driver 2—Non-timber forest production value, Driver 3—Agricultural production value, Driver 4—Livestock production value, Driver 5—Population, State 1—Primary vegetation, State 2—Secondary vegetation, Impact 1—Timber forest production volume, Impact 2—Non-timber forest production, Response—Payments for ecosystem services

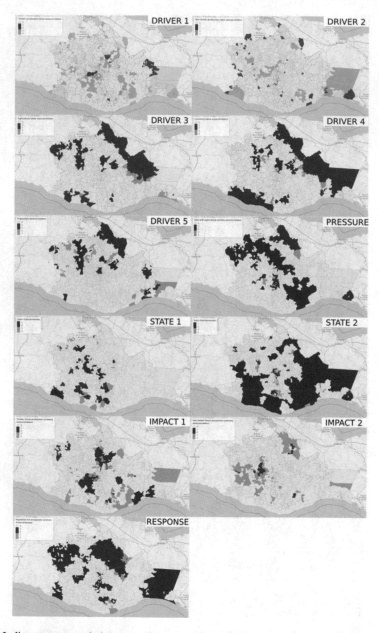

Fig. 3 Indicator autocorrelation maps: Driver 1—Timber forest production value, Driver 2—Non-timber forest production value, Driver 3—Agricultural production value, Driver 4—Livestock production value, Driver 5—Population, State 1—Primary vegetation, State 2—Secondary vegetation, Impact 1—Timber forest production volume, Impact 2—Non-timber forest production, Response—Payments for ecosystem services

Table 3 Results of the local cluster analysis in Oaxaca, Mexico

Indicator	Moran's I	H–H	HH%	L–L	LL%	L-H	LH%	H-L	HL%	N-S	NS%
Timber production (D)	0.020	8	1.4	0	0	48	8.4	3	0.5	511	89.6
Non-timber production (D)	−0.006	0	0	46	8.1	40	7	8	1.4	476	83.5
Main crop value (D)	0.396	29	5.1	88	15.4	11	1.9	0	0	442	77.5
Livestock value (D)	0.356	50	8.8	60	10.5	12	2.1	2	0.4	446	78.2
Population (D)	0.203	35	6.1	68	11.9	20	3.5	5	0.9	442	77.5
Area occupied with cropland (P)	0.467	48	8.4	124	21.8	4	0.7	1	0.2	393	68.9
Primary vegetation (S)	0.137	31	5.4	40	7	20	3.5	1	0.2	478	83.9
Secondary vegetation (S)	0.270	42	7.4	110	19.3	15	2.6	6	1.1	397	69.6
Timber production volume (I)	0.188	31	5.4	5	0.9	49	8.6	6	1.1	479	84
Non-timber production volume (I)	0.041	8	1.4	0	0	32	5.6	9	1.6	521	91.4
Area with PES (R)	0.656	76	13.3	98	17.2	9	1.6	1	0.2	386	67.7

4 Analysis

Oaxaca, Veracruz and Chiapas are areas of intense livestock trading, we assume this could be a reason why livestock production concentrates near borders with these states. It is worth emphasizing that there are areas of high–high concentrations of clusters of livestock and agricultural activity in the north that border with low–low clusters of agricultural and livestock activities. This could indicate positive results of implementing PES in the area and/or transitions into ranching and agricultural activities.

There were very few high–high clusters of primary vegetation and widespread low–low clusters of secondary vegetation. This could be indicative of continuous deforestation in the area and zones where extra efforts should be made to try to maintain the few remaining clusters of primary vegetation and perhaps research what the low–low clusters of secondary vegetation mean. It is quite possible that this is indicative of forest fragmentation.

5 Conclusions

The motivation for developing this geographical information platform is to provide supporting information for decision making regarding forest socio-ecological systems in Oaxaca. Displaying thematic maps of related indicators their autocorrelation of the indicators based on the DPSIR framework can provide a general overview of the state of the forest ecosystem at the municipal level and can be complemented with information collected locally in the 570 municipalities of Oaxaca. Although the fact Oaxaca is subdivided in such small municipalities is complex, it allows better visualization of indicators. Designing the platform with free/open source software and data provided by official government sources as well as data that can be collected by inhabitants of the community is efficient because it requires a small amount of computer storage and processing. Visual exploratory analysis of geospatial data allows cluster analysis of the 570 municipalities of Oaxaca in terms of the DPSIR indicators. The results could contribute to better decision making regarding conservation programs and use of natural resources. Furthermore, local cluster analysis of forestry showed a mix of H–H, L–L, H–L and L–H clusters which is probably an indication of forest fragmentation. We assume that there are areas of low forestry production mixed with areas of high forestry production.

5.1 Recommendations

Continue to provide PES in the zone because they seem to be working as barriers to agricultural and livestock activity expansion. Livestock activities seem to be expanding near borders and coastal regions in the state. We consider that current programs in Mexico such as "Sembrando Vida" can be adjusted in ways that promote combining livestock production activities and productive forestry, often known as agro-forestry in these areas. The results generated a large amount of data that cannot be completely analyzed for this research. We recommend automating an overlay analysis of the results in order to better exploit the amount of data generated.

Appendix

All code is avialable at:
https://github.com/iskarwaluyo/dpsir_autocorrelation_oaxaca_forest

References

Anderies JM, Janssen MA, Ostrom E (2004) A framework to analyze the robustness of social-ecological systems from an institutional perspective. Ecol Soc 9(1)

Anselin L (1995) Local indicators of spatial association-lisa. Geogr Anal 27(93–115):9

Binder CR, Hinkel J, Bots PWG, Pahl-Wostl C (2013) Comparison of frameworks for analyzing social-ecological systems. Ecol Soc 18(4)

Bray DB, Duran E, Ramos VH, Mas JF, Velazquez A, McNab RB, Barry D, Radachowsky J (2008) Tropical deforestation, community forests, and protected areas in the Maya Forest. Ecol Soc 13(2)

Bray D-B, Merino-Pérez L, Barry D (2005) Managing for sustainable landscapes. University of Texas Press, The community forests of Mexico

Burkhard B, Müller F (2008) Driver–Pressure–State–Impact–Response. In: Encyclopedia of ecology. Elsevier, pp 967–970

Cumming G (2014) Theoretical frameworks for the analysis of social-ecological systems. In: Social-ecological systems in transition. Research Institute for Humanity and Nature Springer, Japan, 1 edn., pp 3–27

Ellis AC, Romero Montero EA, Hernández Gómez IU, Anta Fonseca S, López-Paniagua JE (2016) Determinantes de la deforestación en el estado de Oaxaca. Technical report, Agencia de los Estados Unidos para el Desarrollo Internacional (USAID) Proyecto México para la Reducción de Emisiones por deforestación y degradación (M-REDD+), The Nature Conservancy, Rainforest Alliance, Woods Hole Research Center, Espacios Naturales y Desarrollo Sustentable

García Mendoza A, Ordoñez Díaz M, Briones Salas M (2014) Biodiversidad de Oaxaca. Instituto de biología, UNAM : Fondo oaxaqueño para la conservación de la naturaleza: World wildlife fund edition

Hansen MC, Potapov PV, Moore R, Hancher M, Turubanova SA, Tyukavina A, Thau D, Stehman SV, Goetz SJ, Loveland TR, Kommareddy A, Egorov A, Chini L, Justice CO, Townshend JRG (2013) High-resolution global maps of 21st-century forest cover change. Science 342(6160):850–853

Hou Y, Zhou S, Burkhard B, Müller F (2014) Socioeconomic influences on biodiversity, ecosystem services and human well-being: a quantitative application of the DPSIR model in Jiangsu. China. Sci Total Environ 490:1012–1028

Inventarios forestales y de suelos de las Entidades Federativas (2011)

Liu WX, Sun CZ, Zhao MJ, Wu YJ (2019) Application of a DPSIR modeling framework to assess spatial-temporal differences of water poverty in china. JAWRA J Am Water Resour Assoc 55:2

Martin GJ, Camacho Benavides CI, Del Campo García CA, Fonseca SA, Mendoza FC, González Ortíz MA (2011) Indigenous and community conserved areas in Oaxaca. Mexico. Manag Environ Q: Int J 22(2).250–266

Mas JF (2018) Análisis espacial con R: usa R como un sistema de información geográfica. European Scientific Institute Publishing

Moreno IJW, Hernández JG, Gómez AB, Sampayo JCG (2021). Using open-source data and software to analyse land-use changes and deforestation in Marqués de Comillas, Chiapas, Mexico (work in progress). GI_Forum, 1

Nathwani J, Lu X, Wu C, Fu G, Qin X (2019) Quantifying security and resilience of Chinese coastal urban ecosystems. Sci Total Environ 672:51–60

Newton AC, del Castillo RF, Echeverría C, Geneletti D, González-Espinosa M, Malizia LR, Premoli AC, Rey Benayas JM, Smith-Ramírez C, Williams-Linera G (2012) Forest landscape restoration in the drylands of Latin America. Ecol Soc 17(1)

Ostrom E (2009) A general framework for analyzing sustainability of social-ecological systems. Science 325(5939):419–422

Pinto R, de Jonge VN, Neto JM, Domingos T, Marques JC, Patrício J (2013) Towards a DPSIR driven integration of ecological value, water uses and ecosystem services for estuarine systems. Ocean Coast Manag 72:64–79

Ramírez-Marcial N, González-Espinosa M, Musálem-Castillejos K, Eliana Noguera S, Gómez-Pineda E (2014) Estrategias para una construcción social de la restauración forestal en comunidades de la cuenca media y alta del río Grijalva. In: Montañas, pueblos y agua. Dimensiones y realidades de la cuenca Grijalva, pages 518–554. ECOSUR, 1st edn

Roura-Pascual N, Richardson DM, Krug RM, Brown A, Arthur Chapman R, Forsyth GG, Le Maitre DC, Robertson MP, Stafford L, Van Wilgen BW, Wannenburgh A, Wessels N (2009) Ecology and management of alien plant invasions in South African fynbos: accommodating key complexities in objective decision making. Biol Conserv 142(8):1595–1604

Santos-Martín F, Martín-López B, García-Llorente M, Aguado M, Benayas J, Montes C (2013) Unraveling the relationships between ecosystems and human wellbeing in Spain. PLoS ONE 8(9):e73249

Sun C, Wu Y, Zou W, Zhao L, Liu W (2018) A rural water poverty analysis in China using the DPSIR-PLS model. Water Resour Manag 32(6):1933–1951

Svarstad H, Petersen LK, Rothman D, Siepel H, Wätzold F (2008) Discursive biases of the environmental research framework DPSIR. Land Use Policy 25(1):116–125

Tsai HT, Tzeng SY, Fu HH, Wu JC (2009) Managing multinational sustainable development in the european union based on the DPSIR framework. Afr J Bus Manag 3:727–735

Vacik H, Wolfslehner B, Seidl R, Lexer MJ (2007) Integrating the DPSIR approach and the analytic network process for the assessment of forest management strategies. In: Reynolds KM, Thomson AJ, Köhl M, Shannon MA, Ray D, Rennolls K (eds) Sustainable forestry: from monitoring and modelling to knowledge management and policy science. CABI, Wallingford, pp 393–411

Velázquez A, Durán E, Ramriez I, Mas J-F, Bocco G, Ramriez G, Palacio J-L (2003) Land use-cover change processes in highly biodiverse areas: the case of Oaxaca. Mexico. Global Env Change 13(3):175–184

Vidal-Abarca MR, Suárez-Alonso ML, Santos-Martín F, Martín-López B, Benayas J, Montes C (2014) Understanding complex links between fluvial ecosystems and social indicators in Spain: an ecosystem services approach. Ecol Complex 20:1–10

Wei Y, Zhu X, Li Y, Yao T, Tao Y (2019) Influential factors of national and regional CO2 emission in China based on combined model of DPSIR and PLS-SEM. J Clean Prod 212:698–712

Algorithms and Methods for Geospatial Data

Fourier Transform Based Methods for Unwrapping of Sentinel-1 Interferograms

Alejandro Téllez-Quiñones, Juan Carlos Valdiviezo-Navarro, and Alejandra A. López-Caloca

Abstract Synthetic aperture radar (SAR) applications related to phase estimation, just as studies on subsidence, generation of digital elevation models or time series analyses of Earth surface changes, require an adequate phase unwrapping (PU) process. This process should imply at least a qualitative fringe analysis of certain small regions of the complete interferogram. Although, this qualitative analysis is usually omitted or cannot be followed due to its complexity, sometimes the form of the observed fringes allows it. In general, the form of the estimated phase depends on the properties or capabilities of the PU algorithm used, as well as its input parameters, and these points may be unknown for the user. For instance, in applications with Sentinel-1 images, we have an algorithm called SNAPHU, which is quite robust and effective. Nevertheless, sometimes, it produces inconsistent results according to the theoretical form of the fringes and the expected form of the phase. Consequently, the proposal to solve these failed results could involve implementing different PU routines, that can be more familiar or relatively easy-to-program, as the fast Fourier transform (FFT) based method. In this manuscript, we describe a simple way to do this analysis, considering a small subset of a SAR deformation interferogram, and three particular PU algorithms; SNAPHU, the FFT based method, and a proposal or extension of this last one, but based on the use of continuous Fourier transforms.

1 Introduction

Phase unwrapping algorithms are strategies to demodulate phase signals in many physical and engineering applications, particularly, the design of these algorithms

A. Téllez-Quiñones (✉) · J. C. Valdiviezo-Navarro
CONACYT-CentroGeo, Sierra Papacal-Chuburná Km 5, 97302 Mérida, Yucatán, México
e-mail: atellez@centrogeo.edu.mx

J. C. Valdiviezo-Navarro
e-mail: jvaldiviezo@centrogeo.edu.mx

A. A. López-Caloca
CentroGeo-CONACYT, Contoy 137, 14240 Tlalpan, Ciudad de México, México
e-mail: alopez@centrogeo.edu.mx

© The Author(s), under exclusive license to Springer Nature Switzerland AG 2022
R. Tapia-McClung et al. (eds.), *Advances in Geospatial Data Science*, Lecture Notes in Geoinformation and Cartography, https://doi.org/10.1007/978-3-030-98096-2_6

represents a key point for SAR interferometry (Ghiglia and Pritt 1998; Hanssen 2001). A SAR image is represented by the complex model $\mu(x, r) = A(x, r)e^{i\varphi(x,r)}$, where i is the imaginary unit, and the phase term involved is given by a real-valued scalar function $\varphi = \varphi(x, r)$ which is in terms of coordinates x and r, called azimuth and range, respectively. The phase term represents the double path travel of microwave radiation between the object (terrain surface) and the SAR sensor (satellite platform), whereas the amplitude term $A = A(x, r)$ is another real-valued function (non-negative) that represents physical terrain features related to the kind of structures or materials in the imaged object surface. In a digital context a SAR image is described as $\mu(m, n) = \mu(x_m, r_n)$, as well as its amplitude and phase terms, by considering pixel positions (m, n) with $m, n = 1, 2, \ldots, M, N$, respectively. These positions are related to a discrete sampling $x_m = x^- + (m - 1)\Delta x, r_n = r^- + (n - 1)\Delta r$, where x^- and r^- are reference constants, and Δx and Δr are system resolution parameters (Téllez-Quiñones et al. 2020). Therefore, when considering two co-registered SAR images of the same geographical zone $\mu_1 = A_1 e^{i\varphi_1}$ and $\mu_2 = A_2 e^{i\varphi_2}$, an *interferometric signal*

$$s(m, n) := \frac{\sum \mu_1(p, q)\mu_2^*(p, q)}{\sqrt{\sum |\mu_1(p, q)|^2}\sqrt{\sum |\mu_2(p, q)|^2}} \approx \gamma(m, n)e^{i\phi(m,n)}, \qquad (1)$$

can be obtained. In this case, $\sum = \sum_{(p,q)\in\mathcal{W}_{m,n}}$, $\mathcal{W}_{m,n}$ is a window of pixels centered at (m, n), $*$ denotes complex conjugation, $\gamma := |s|$ is the complex module of s (a term called *coherence*), and $\phi := \varphi_1 - \varphi_2 \approx \phi_{topo} + \phi_{defo}$ is the so-called *interferometric phase*, a function that can be decomposed in many other phase terms like a *topographical phase* ϕ_{topo} and a *deformation phase* ϕ_{defo} (assuming ground subsidence or elevation due to seismic activity). Thus, the signal s may be redefined by considering reference terms ϕ_{topo}^{ref}, based on reference digital elevation models (DEMs) of the zone, in order to obtain

$$s_{defo} := se^{-i\phi_{topo}^{ref}} \approx \gamma e^{i\phi_{defo}}. \qquad (2)$$

In other words, we can obtain a signal s_{defo} that contains the information of the phase term of interest, in this case, a deformation phase $\phi_{defo} := (4\pi/\lambda)|AB'|$, as described in Fig. 1. In this figure, P_1 is the position where the *primary image* μ_1, was captured, and the information of a point A is detected through the line of sight (LOS) $|P_1A|$. After certain revisit time of the satellite platform (six days for platforms Sentinel-1 A and B), the sensor captures the information of the same point from position P_2, but the point A may be displaced some few centimeters to position B due to a quake. Therefore, the *secondary image* μ_2 possess the information of point B through the line $|P_2B|$ intersected in point C with LOS $|P_1A|$ and the deformation phase is proportional to the distance with sign $|AB'|$, where B' is the orthogonal projection of B to the LOS $|P_1C|$. So, a demodulation process is required to extract the information of $\phi = \phi_{defo}$ by considering the signal $s_\phi = \gamma e^{i\phi}$ and the wrapped phase

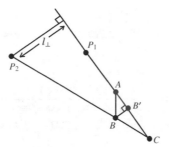

Fig. 1 Geometrical displacement in LOS $|AB'|$ defined by $\phi_{defo} = (4\pi/\lambda)|AB'|$, where λ is the wavelength. Distance $|AB'|$ is expected to be at wavelength orders (some few centimeters) in comparison with distances $|P_1 A|$ and $|P_2 B|$ (many kilometers). The orthogonal baseline $l_\perp$ (some few meters), is also exemplified

$$\psi = W[\phi] := \text{atan2}[\text{Im}(s_\phi), \text{Re}(s_\phi)], \qquad (3)$$

where W is called *wrapping operator* (Ghiglia and Pritt 1998), atan2$[\cdot, \cdot]$ is the extended inverse tangent with range $(-\pi, \pi]$, $\text{Im}(s_\phi) := \gamma \sin(\phi)$, and $\text{Re}(s_\phi) := \gamma \cos(\phi)$.

In SAR interferometry, a wrapped phase ψ represents an *interferogram*, a fringe pattern given by a discontinuous version of ϕ with 2π-jumps or discontinuities induced by the wrapping operation $\psi = W[\phi] = \phi + 2\pi\eta$, where η is an unknown integer distribution that "wraps" the phase ϕ, i.e. $-\pi < \psi \leq \pi$. Thus, the main goal of any unwrapping algorithm is to remove these discontinuities, in order to estimate ϕ, with certain smoothness in some cases, from the information of ψ. This can be achieved by many strategies that conform an extensive field of study (Chen and Zebker 2000; Constantini 1998; Flynn 1997; Goldstein et al. 1988; Lyuboshenko and Maître 1999; Zebker and Goldstein 1986), however, not all of these routines are easy to implement or successfully applied (Téllez-Quiñones et al. 2020).

2 Proposed Method

To be concise in notation with Ghiglia and Pritt (1998), let us assume $r = y$ in order to describe the scalar field $\phi = \phi(x, y)$ in terms of continuous coordinates (x, y) as usual. Thus, an image or an $M \times N$-array interpretation of this scalar field is $\phi(m, n) = \phi(x_m, y_n)$ with $m, n = 1, 2, \ldots, M, N$. According to Ghiglia and Pritt (1998), a PU problem can be translated in solving a discrete Poisson equation $\nabla^2 \phi(m, n) = \rho(m, n)$ for ϕ, where ρ is calculated as $\rho(m, n) := (\Delta^x_{m,n} - \Delta^x_{m-1,n}) + (\Delta^y_{m,n} - \Delta^y_{m,n-1})$ and $\Delta^x_{m,n} := W[\psi(m + 1, n) - \psi(m, n)]$, $\Delta^y_{m,n} = W[\psi(m, n + 1) - \psi(m, n)]$. In this case, the discrete Laplacian is given by $\nabla^2 \phi(m, n) = (\phi(m + 1, n) - 2\phi(m, n) + \phi(m-1, n)) + (\phi(m, n+1) - 2\phi(m, n) + \phi(m, n - 1))$. So, by considering extended periodic versions of ϕ and ρ, i.e. the

mirror reflections or $(2M - 1) \times (2N - 1)$-arrays $\tilde{\phi}$ and $\tilde{\rho}$, respectively, the discrete
Fourier transforms or fast Fourier transforms (FFTs) $\Phi = \mathrm{FFT}[\tilde{\phi}]$ and $P = \mathrm{FFT}[\tilde{\rho}]$
are related by

$$\Phi(p, q) = \frac{P(p, q)}{2\cos(\pi p/M) + 2\cos(\pi q/N) - 4},$$
$$p, q = 1, \ldots, (2M - 1), (2N - 1). \tag{4}$$

Hence, an FFT based method for PU is implied by the calculation of the dis-
crete inverse formula $\tilde{\phi} = \mathrm{IFFT}[\Phi]$ and the fact that $\phi(m, n) = \tilde{\phi}(m, n)$ for $m, n =$
$1, \ldots, M, N$.

On the other hand, considering a reference continuous domain $\Omega = [-L_x, L_x] \times$
$[-L_y, L_y]$, we can also obtain a continuous formulation of this problem, by
solving the continuous version of the negative Poisson equation $-[(\partial^2/\partial x^2) +$
$(\partial^2/\partial y^2)]\phi = h$. In this formulation, we just need to consider appropriate scaling factors $\Delta_x = 2L_x/(M - 1)$ and $\Delta_y = 2L_y/(N - 1)$ for estimating first and
second continuous partial derivatives related to h and approximating with simple Riemann sums the continuous Fourier transform (CFT) of a scalar func-
tion $f = f(x, y)$ given by $F(u, v) = \mathscr{F}[f(x, y)] := \int_{\mathbb{R}^2} f(x, y)e^{-i2\pi(ux+vy)}dxdy$.
In this case, it is assumed that $f(x, y) = 0$ for all $(x, y) \notin \Omega$, and an ideal
frequency domain $(u, v) \in [-U, U] \times [-V, V]$ with $U = (M - 1)/(4L_x)$, $V =$
$(N - 1)/(4L_y)$, as proposed in Bertero and Boccacci (1998). A similar procedure can be applied with the inverse expression $f(x, y) = \mathscr{F}^{-1}[F(u, v)] :=$
$\int_{\mathbb{R}^2} F(u, v)e^{+i2\pi(ux+vy)}dudv$, by considering scaling factors $\Delta_u = (2U)/(M - 1)$
and $\Delta_v = (2V)/(N - 1)$. Therefore, from these formulas and Fourier transform
properties, we can find the solution in frequency space, as $\Phi = H/K$ with $\Phi(u, v) =$
$\mathscr{F}[\phi(x, y)]$, $H(u, v) = \mathscr{F}[h(x, y)]$ and $K(u, v) = 4\pi^2(u^2 + v^2)$, i.e. $\phi(x, y) =$
$\mathscr{F}^{-1}[\Phi(u, v)]$. Of course, in this approach we first calculate the negative Laplacian
$h(m, n) = h(x_m, y_n) = -[(\Delta_{m,n}^x - \Delta_{m-1,n}^x)/(\Delta_x^2)] - [(\Delta_{m,n}^y - \Delta_{m,n-1}^y)/(\Delta_y^2)]$ and
then the continuous spectra $H(p, q) = H(u_p, v_q) = \mathscr{F}[h]$ for finally estimating
the phase spectra $\Phi(p, q) = \Phi(u_p, v_q)$ with $(u_p, v_q) \in [-U, U] \times [-V, V]$, by
considering regular partitions $x_m = -L_x + (m - 1)\Delta_x$, $y_n = -L_y + (m - 1)\Delta_y$,
$u_p = -U + (p - 1)\Delta_u$, $v_q = -V + (q - 1)\Delta_v$, with $m, p = 1, \ldots, M$ and $n, q =$
$1, \ldots, N$, respectively.

The Riemann-sum implemented in our algorithm to calculate a CFT of array
$f(m, n) = f(x_m, y_n)$ is

$$F(p, q) = \sum_{m=1}^{M-1} \sum_{n=1}^{N-1} f(m, n)e^{-i2\pi(u_p x_m + v_q y_n)} \Delta_x \Delta_y, \quad \forall p, q = 1, \ldots, M, N, \tag{5}$$

considering a fictitious reference domain $\Omega = [-1/2, 1/2] \times [-1/2, 1/2]$, and a
similar formula is applied for the inverse operation $f(m, n) = \mathscr{F}^{-1}[F(p, q)]$. Nevertheless, in the calculation of wrapped partial derivatives from array $\psi(m, n)$, a

zero-padding operation (2-pixel width border of zeros added to the original array) in the wrapped phase is used, in order to warranty a band-limited spectra $\Phi(p, q)$ (Bertero and Boccacci 1998), which implies the validity of Eq. (5) for Φ. Then, a mirror reflection operation is applied to the zero padded wrapped phase to estimate a periodic extension of array $h(m, n)$, in order to warranty the existence (relative) of the Fourier series representation $\phi(m, n) = \mathscr{F}^{-1}[\Phi(p, q)]$ (properly, for its periodic extension $\tilde{\phi}$). Finally, we have tested the functionality of these two algorithms by considering many synthetic examples similar to those described in Téllez-Quiñones et al. (2019).

3   Region of Analysis

The Icelandic volcano Fagradalsfjall is located in the Reykjanes Peninsula in southwest Iceland. This volcano is in the Krýsuvík-Trölladyngja volcanic system. It is a large volcano formed by layers from successive eruptions. From a geological standpoint, it is located in a zone of active rifting, at the divergent boundary between the Eurasian and North American plates. The mountain range complex of the Reykjanes Peninsula, where mount Fagradalsfjall is located, represents one of the zones where earthquakes and seismic displacements are continuously produced due to the presence of its active volcanoes. Volcanic eruptions naturally record the ascent (of the volcano's surface, causing it to rise) and subsidence of the surfaces of the volcanic mountains caused by their activity. These types of places require the monitoring and follow-up of volcanic activity (Biggs and Wright 2020). The use of interferograms and displacement maps show the volcanic activity of the site as shown in Figs. 2 and 3. When observing interference fringes, colored bands caused by earthquakes on the surface of the volcano during the volcanic eruption are observed. These changes are due to the characteristics of the eruption and the properties of the soil that cause the interference fringes to move differently. Each cycle of interference color represents ground movement in LOS direction in units of the radar wavelength (2.8 cm for Sentinel-1) between color cycles (Massonnet et al. 1994). The most important volcano-tectonic activities during 2021 were registered in February, March, May, and June, and the volcano is still spewing and expanding its lava flow field. A review of this tectonic activity has been reported in Cubuk-Sabuncu et al. (2021). Due to the seismic features of this region, we have used Sentinel-1 images during March period for our PU experiment in the recovery of an estimated ϕ_{defo}.

3.1   Materials and Methods

We have obtained an interferogram subset in gray-scale levels as in Fig. 3 by following first the generation of the complete Iceland interferogram in false color (rainbow-colored) in the same figure. This interferogram was generated in the Sentinel Appli-

Fig. 2 Deformation SAR interferogram recovered with Sentinel-1 images captured at March 07 and 19, 2021. The terrain-corrected interferogram is displayed in a region of Iceland

Table 1 Characteristics of IW-SLC products used to generate the interferogram ψ to be processed: Acquisition date (A.D.), orthogonal baseline $l_\perp$ (in meters), mean coherence $\bar{\gamma}$, platform (Pl), polarization (Po), reference DEM, and pass mode (P.M.)

A.D.	$l_\perp$	$\bar{\gamma}$	Pl	Po	DEM	P.M.
03/07/2021			B	VV		Descending
	27.95	0.6233			Copernicus (30 m)	
03/19/2021			B	VV		Descending

cation Platform (SNAP) free software by implementing the basic steps described in Braun and Veci (2020). These steps were applied to two interferometric wide swath-single look complex (IW-SLC) Sentinel-1 images of Iceland which features are displayed in Table 1.

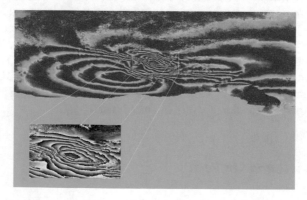

Fig. 3 Region subset of 553 × 343 pixels taken from the Iceland interferogram. The subset has an interesting lobular zone with relative high frequency fringes that theoretically should be for a convex or a concave deformation surface induced by the seismic activity

Between these steps we found the *split operation* and the application of *orbit files* to both images. Then, we need to apply the *coregistration of the image pair*, the *interferogram generation*, a *deburst operation* and an *interferogram filter*. In the case of split operation, it corresponds to an election of subregions in the complete image. The SAR image products may contain real, imaginary and intensity parts of model μ, in georeferenced pixel-arrays for different zones named IW1, IW2 and IW3. Each IW-zone is also divided in subregions called bursts (at about 9 subregions), that can be selected by the split operation. Thus, for our experiments we selected IW2-bursts 7 and 8, respectively. Additionally, the application of orbit files, a refreshing of geometrical parameters of orbits where the images were captured, should be conducted. In our case, we specified a Copernicus 30 m resolution DEM, for the coregistration process.

Once the images were coregistered, the interferogram formation induced by Eqs. (1)–(3) is applied, and the subregion division given by the selected bursts is removed by the deburst operation. After the interferogram formation, the Goldstein wrapped phase filter (Goldstein and Werner 1998) was applied, and the final deformation interferogram ψ is displayed in Fig. 3. Finally, we took a subset of this wrapped phase with $M \times N = 553 \times 343$ pixels (gray-scale image in Figs. 3 and 4a), for a simple local fringe analysis, respectively. When the wrapped phase ψ is ready, it can be unwrapped by a standard tool called statistical-cost network-flow algorithm for phase unwrapping (SNAPHU), as described in Chen and Zebker (2000), by following the steps in Braun and Veci (2020). Nevertheless, we can also unwrap this real data by considering own-code implementations in Scilab 6.0.2 free software of alternative and simple PU algorithms, like the FFT based method and the CFT proposal, previously described. To do this, we can export the information of the interferogram subset into a CSV file, by considering the menu options in SNAP. Then, with some code instructions in a sci-file (Scilab file), we can read the CSV file, first as a string-array converted to a double precision vector, and then converted to an $M \times N$-matrix array or variable, that can be saved in a sci-data file. This data can be read and unwrapped by another sci-file that contains code instructions for the corresponding PU algorithm.

Since the FFT based method (and also our CFT proposal) is considered as an L^2-norm algorithm or a least-squares solution (Ghiglia and Pritt 1998), the unwrapped result has minimum gradient variation according to the square norm or potential

$$ J[\phi] = \sum_{m,n} [(\phi(m+1,n) - \phi(m,n) - \Delta_{m,n}^x)^2 + (\phi(m,n+1) - \phi(m,n) - \Delta_{m,n}^x)^2]. $$

(6)

Thus, an estimated phase ϕ_0 from such an algorithm represents a good match in terms of phase derivatives, that does not necessarily satisfy a re-wrapped test, i.e. $W[\phi_0] \neq \psi$. When this happens, a post-processing operation called *congruence* can be considered. This operation is given by $\phi_1 = \phi_0 + W[\psi - \phi_0]$, and the result is such that $W[\phi_1] \cong \psi$. Although this operation may be useful, as described in Pritt (1997), it can introduce some unexpected 2π-discontinuities (inconsistencies) in the result that may induce a non-smooth phase estimate ϕ_1.

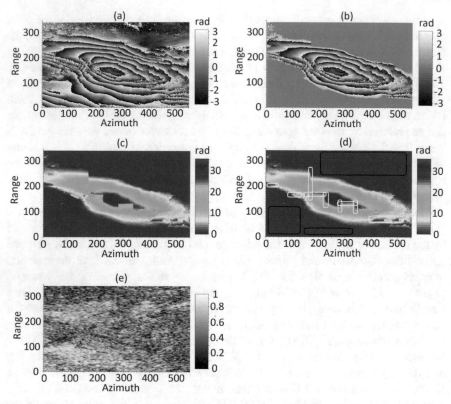

Fig. 4 **a** Original subset of the wrapped phase. **b** Re-wrapped SNAPHU estimation. **c** Unwrapped phase by SNAPHU. **d** Inconsistencies in the unwrapped phase in **c**, enclosed by black and white lines. **e** Coherence map γ (dimensionless and with range $0 \leq \gamma \leq 1$, by construction)

Once the unwrapped estimation is obtained in Scilab, the $M \times N$-array of the phase ϕ_0 (or ϕ_1) can be converted into a double precision vector, then into a string array, and finally the data of this last one array can be written as a CSV file similar to the exported file. Therefore, returning to SNAP software, the CSV file with unwrapped data can also be imported to some product previously created with the original data in SNAP. This last product could be the original interferogram subset with its own georeferenced pixels in longitude and latitude, respectively.

3.2 Results

We have unwrapped the interferogram subset with 3 PU algorithms; the SNAPHU routine implemented in C-code, and the FFT and CFT methods implemented in Scilab code. The SNAPHU unwrapped result can be shown in Fig. 4c. In the case of

SNAPHU estimation, we can see in Fig. 4d that the result has some inconsistencies; straight lines of discontinuities enclosed in white lines and unexpected regions of zeros enclosed in black lines. The last ones are inconsistencies because theoretically, and according to the form of the fringes in the interferogram, the expected estimate of ϕ on the complete domain should be some kind of relative smooth concave or convex surface for circular or elliptical fringes. Moreover, those regions where the fringes are "thick" and "open" should represent a smooth slope surface, which is not necessarily a constant. As described in Téllez-Quiñones et al. (2020), a possible explanation of this failed result may be given due to the strong dependence of this algorithm to the coherence γ (see Fig. 4e) in the minimization of the particular potential function $J = J_{SNA}[\phi]$ (different from J in Eq. (6)) that describes the SNAPHU algorithm. However, in this case, the unwrapping process could be achieved without considering the coherence map γ for masking or penalizing some areas, as SNAPHU algorithm has done. This is because the mean coherence $\bar{\gamma} = 0.4972$ for the map in Fig. 4e, is quite good and, according to the fringes in the subset, they look without too much amount of noise (ψ in Fig. 4a has an approximated signal-to-noise-ratio of $SNR = 13.45$dB).

In comparison with FFT and CFT based methods, the re-wrapped results with congruence operation added (displayed in Fig. 5), look better than the re-wrapped phase in Fig. 4b, which is only good, outside the mask of zeros produced by the SNAPHU estimation. To have a quantitative idea of the accuracy in wrapped phase, achieved by these methods, we have calculated the root-mean-square errors $RMSE = [(1/[MN]) \sum_{m,n} (\psi(m, n) - \psi_{Met}(m, n))^2]^{1/2}$, where ψ is the current wrapped phase, $\psi_{Met} = W[\phi_{Met}]$, and ϕ_{Met} is an unwrapped phase obtained by some of the algorithms or methods previously described. These errors are displayed in Table 2.

4 Discussion and Conclusions

Although the results in re-wrapped phases are good for our FFT and CFT estimations with post-processing of congruence, we must consider an expected unwrapped phase with certain degree of smoothness in all its domain, therefore we have plotted these estimations in Fig. 6 for a qualitative comparison. We also made other reconstructions in Scilab with algorithms based on local polynomial models (Téllez-Quiñones et al. 2019), but the results were unsatisfactory because we have not yet implemented and adequate path-following strategy (Ghiglia and Pritt 1998) to robust them. Nevertheless, independently of the method used, in this manuscript we have described a simple way to process relatively small data arrays of SAR interferograms, considering simple unwrapping algorithms implemented in Scilab. The dimensions of these arrays depend on the capabilities of the CSV files and the SNAP export tool to store the information into this format, which in the practice is not much (arrays smaller than 500×500 pixels). However, the amount of information that can be unwrapped could increase when considering other types of exported files, such as GeoTIFF, GDAL, etc. It is important to note that although the best expected result

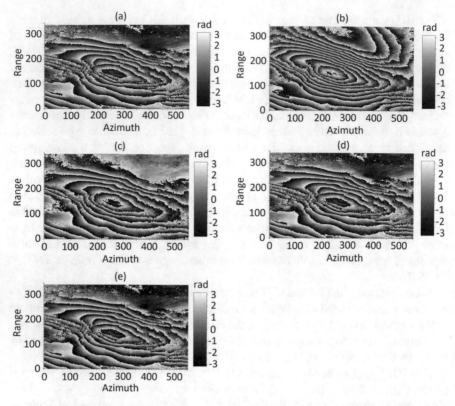

Fig. 5 **a** Original wrapped phase, and re-wrapped phases; **b** FFT based method, **c** proposed CFT based method, **d** FFT based method with congruence post-processing, and **e** the proposed method with congruence post-processing

Table 2 Root mean square errors in [rad] for the re-wrapped phases. Post-processing operation of congruence (cong.) has been considered in our estimations*

Method	RMSE	Method	RMSE	Method	RMSE
SNAPHU	1.3579235	FFT	2.5896342	CFT	2.5656525
		FFT cong.	1.026D-15	CFT cong.	7.274D-16

*SNAPHU estimation does not require this operation

was achieved with the CFT method with congruence, in this particular experiment, we must not assume that the aforementioned algorithm is better than SNAPHU in any other case. Actually, the key point to improve the SNAPHU estimation depends on the knowledge and manipulation of its internal parameters, something that, unfortunately, is not feasible for many users since its coding is quite extensive and complex. In absence of this knowledge, we have proposed the use of algorithms based on Fourier transforms, which could be useful alternatives, in situations such as we have analyzed in this work.

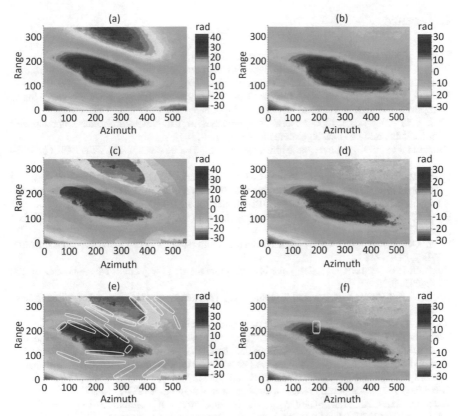

Fig. 6 Unwrapped phases; **a** FFT based method, **b** proposed CFT based method, **c** FFT based method with congruence post-processing, and **d** the CFT based method with congruence post-processing. Inconsistencies in the congruent estimations; **e** FFT based method, **f** CFT based method

Finally, although we implemented formula in Eq. (5) to calculate a CFT in a general way, it can be considered the implementation of shifting and FFT operations (with sci-functions) to calculate a CFT (See page 41 in Bertero and Boccacci 1998). Thus, the FFT implementation reduces the computational complexity of a DFT from $O(\omega^2)$ to $O(\omega \log(\omega))$ where $\omega = M \times N$. Nevertheless, for a simple comparison of our proposal with the basic DFT formula, we could say that it has an approximated complexity upper bound given by $O(\omega^2 + 3\omega)$, considering the fact that a CFT can be obtained from two shifting operations, a DFT operation, and a multiplication by a scaling factor.

References

Bertero M, Boccacci P (1998) Introduction to inverse problems in imaging. IOP, London

Biggs J, Wright TJ (2020) How satellite InSAR has grown from opportunistic science to routine monitoring over the last decade. Nat Commun. https://doi.org/10.1038/s41467-020-17587-6

Braun A, Veci L (2020) Sentinel-1 Toolbox: TOPS Interferometry Tutorial. ESA: SkyWatch Space Applications Inc. http://step.esa.int/main/doc/tutorials

Chen CW, Zebker HA (2000) Network approaches to two-dimensional phase unwrapping: intractability and two new algorithms. JOSA A 17:401–414

Constantini M (1998) A novel phase unwrapping method based on network. IEEE TGRS 36:813–821

Cubuk-Sabuncu Y, Jónsdóttir K, Caudron C, Lecocq T, Parks MM, Geirsson H, Mordret A (2021) Temporal seismic velocity changes during the 2020 rapid inflation at Mt. Thorbjorn-Svartsengi, Iceland, using seismic ambient noise. GRL. https://doi.org/10.1029/2020GL092265

Flynn TJ (1997) Two-dimensional phase unwrapping with minimum weighted discontinuity. JOSA A 14:2692–2701

Ghiglia DC, Pritt MD (1998) Two-dimensional phase unwrapping: theory, algorithms, and software. Wiley, New York

Goldstein RM, Werner CL (1998) Radar interferogram filtering for geophysical applications. GRL 25:4035–4038

Goldstein RM, Zebker HA, Werner CL (1988) Satellite radar interferometry: two-dimensional phase unwrapping. Radio Sci 23:713–720

Hanssen RF (2001) Radar interferometry: data interpretation and error analysis. KAP, NY

Lyuboshenko I, Maître H (1999) Phase unwrapping for interferometric synthetic aperture radar by use of Helmholtz equation eigenfunctions and the first Green's identity. JOSA A 16:378–395

Massonnet D, Feigl K, Rossi M, Frédéric A (1994) Radar interferometric mapping of deformation in the year after the Landers earthquake. Nature 369:227–230

Pritt MD (1997) Congruence in least-squares phase unwrapping. IEEE IGARSS Proc. 2:875–877

Téllez-Quiñones A, Salazar-Garibay A, Valdiviezo-Navarro JC, Hernandez-Lopez F, Silván-Cárdenas JL (2020) DInSAR method applied to dual-pair interferograms with Sentinel-1 data: a study case on inconsistent unwrapping outputs. IJRS 41:4662–4681

Téllez-Quiñones A, Legarda-Sáenz R, Salazar-Garibay A, Valdiviezo-NJC, León-Rodrígez M (2019) Direct phase unwrapping method based on a local third-order polynomial fit. Appl Opt 58:436–445

Zebker HA, Goldstein RM (1986) Topographic mapping from interferometric synthetic aperture radar observations. JGR 91:4993–4999

Methodology and Relevant Results About an Area-Based Conservation Indicator of Superficial Water Bodies for the Grijalva Basin

Felipe Omar Tapia-Silva ⓘ, **Aymara O. Ramírez-González** ⓘ, and **José Luis López-Gonzaga** ⓘ

Abstract Superficial water bodies play a fundamental role in ecological and socioeconomic terms. Therefore, it is necessary to investigate methods to know which may be losing their persistence over time. In this manuscript, the methodology and main results of the spatial analysis process utilized for generating an area-based conservation indicator are presented. Binary monthly water-no-water layers were obtained from the processing of SAR Sentinel 1a images. A spatial layer of water bodies built with data from three decades served as a reference to obtain the indicators of change with respect to the area registered in this multi-year layer. Considering that the monthly indicator was calculated as a percentage of its area in the multiannual reference layer, the obtained results can be defined as long-term change indicators. It is concluded that some water bodies considered permanent in the multiannual reference layer did not cover, during the study period (2016–2018), the values of the area recorded in that layer, but that most remained close to those values. A 30% reduction in the area covered by these bodies in the reference layer was observed. Regarding temporary water bodies, these were detected with a reduction of more than 80% as compared to the value in the reference layer, which may be an indicator of their gradual loss process. The information generated can be used for management and conservation purposes at the regional and local levels of water bodies.

1 Introduction

Superficial water bodies can be defined as areas covered with water, with or without vegetation, and that are permanently (PW) or temporary (TW) (Yamazaki et al. 2015). Considering that they play a fundamental role in ecological and socioeconomic terms,

F. O. Tapia-Silva (✉) · A. O. Ramírez-González
UAM-Iztapalapa, Mexico City, Mexico
e-mail: ftapia@izt.uam.mx

J. L. López-Gonzaga
CentroGeo, CDMX, Mexico City, Mexico
e-mail: jllopez@centrogeo.edu.mx

it is necessary to investigate methods that allow knowing which ones may be in the process of losing their persistence over time with the aim to conserve them.

Water bodies and wetlands in tropical areas have received much less scientific attention than other ecosystems such as tropical forests and other cover types in non-flooded areas (Ellison 2004). In addition, there is little basic information on its extension and what does exist is not very accessible to researchers, managers, and designers of public policies (Ellison 2004). Few research reports have been published for the basin of interest, such as those of Saud et al. (2018) and (Sánchez et al. 2015) that essentially describe the ecological role and the provision of environmental services that the water bodies in the area can offer.

SAR images are very useful for detecting surfaces in areas with frequent cloudiness, such as the study area, which is the Grijalva Basin, particularly in its lower part (called the Bajo Grijalva) with the highest water bodies concentration in the basin (see Fig. 2).

A reference layer (RL) is essential as an element of comparison to detect changes that the water bodies in some regions may be undergoing in a certain period and to define their persistence in terms of the area covered with water. The RL should serve as a reference for the stable or expected condition. It is necessary that this layer has been obtained through a clear, replicable, and scientifically reliable methodology and based on multi-temporal high spatial resolution data.

The purpose of this article is to present and discuss the methodology and some of the most relevant results to obtain an area-based indicator related to the state of conservation of superficial water bodies located in the Mexican part of the Grijalva Basin. The indicator is the monthly area covered with water by each water body expressed as a percentage of its value in the RL. This layer defines the expected state of each water body during the last three decades. Considering that the monthly indicator was calculated as a percentage of its value in the multi-year RL, the obtained results can be defined as long-term change indicators.

The study was carried out in the context of an investigation to monitor territorial processes using satellite images at the regional level in Mexico.[1]

2   Methodology

The methodology developed is shown in Fig. 1. It consists of three groups of activities: image processing, obtaining the RL and generating indicators. These are detailed below.

---

[1] Mentioned in the acknowledgements.

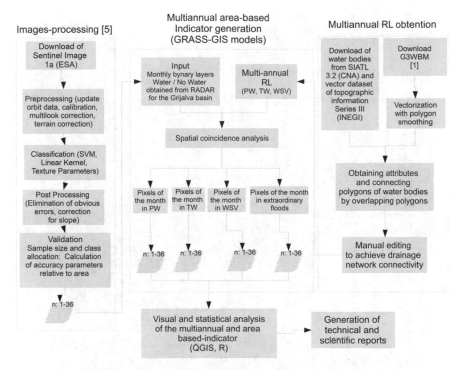

Fig. 1 Methodology developed for generation and visualization of the multiannual area-based indicator related to the Reference Layer (RL). PW: Permanent bodies, TW: temporary bodies and WSV: wet soils and vegetation

2.1 Image Processing

Radar images allowed the generation of monthly maps from 2016 to 2018 delineating the areas covered by water. As pointed out in López-Caloca et al. (2020), 181 S1 C-band images were obtained from Level-1 Ground Range Detected (GRD) S1 SAR and downloaded from the Copernicus Open Access Hub (https://scihub.copern icus.eu/dhus/#/home). VV polarization data were used. The detailed methodology of processing and validation of the SAR Sentinel 1a images can be consulted in López-Caloca et al. (2020). In short, this consisted of the following: (a) Image preprocessing, which included updating of orbit data, calibration, and multi look and terrain corrections; (b) Obtaining the water-not water binary masks using support vector machines. After that, post-processing activities were carried out to eliminate obvious errors and to correct for slope. (c) Validation of the classification results. The performed activities were calculation of the sample size, assignment of value to the binary classes and sampling randomization according to Olofsson et al. (2014), and finally, calculating of accuracy parameters relative to the area following that proposed by Pontius and Santacruz (2014). The obtained accuracy levels were close to 90% (López-Caloca et al. 2020) for the water-no water binary layers classification.

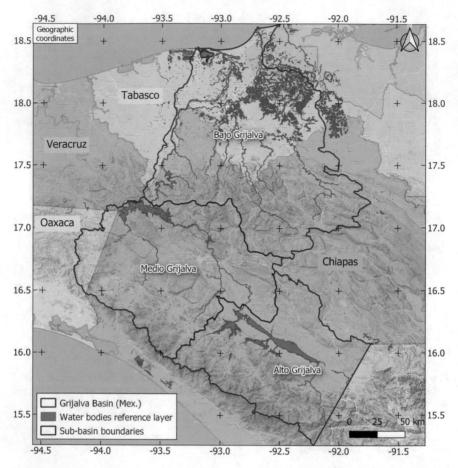

Fig. 2 Obtained multianual reference layer (RL) of water bodies in the Grijalva Basin in Mexico

2.2 Multiannual RL Obtention

Global ~90 m Water Body Map (G3WBM) (Yamazaki et al. 2015) is the multi-year product of surface water bodies selected as RL that classifies water bodies in PW and TW. Additionally, it includes wet soils and vegetation (WSV). G3WBM had to be processed to ensure connectivity between rivers and water bodies in the basin and to include the descriptive component as written in the following. G3WBM data was downloaded in raster format from http://hydro.iis.u-tokyo.ac.jp/~yamadai/G3WBM/ upon request for a password. Afterward, the downloaded raster layer was converted to vector taking the categories of each type of water body as the value for each pixel. The categories are WSV: 20, TW: 40, and PW: 50 and 51. This procedure was performed in ArcGIS 10.2. In this way, a polygons vector file with squared contours of 90 × 90 m (corresponding to the resolution of the G3WBM raster) was obtained. To smooth the resulting polygons, the algorithms of ArcGis Simplify Building, Simplify Polygon, and Smooth Polygon were tested. The one selected was SmoothPolygon, with a tolerance of 50 m, and the Paek method, considering that it has the advantage of smoothing the contours avoiding losing smaller areas.

The next step was to ensure continuity in the network of rivers and water bodies. The missing areas were added based on the 1:50,000 layer of water bodies of the SIATL 3.2 of Mexican INEGI, available in the Watershed Water Flow Simulator (SIATL, version 3.2). The guidelines to carry out this activity were: (a) To maintain the polygons derived from G3WBM and (b) To connect non-linear water bodies (lakes, ponds, dams) present in G3WBM. To assign an identifier and name to the water bodies, the INEGI series III vector data set of topographic information, scale 1:50,000, was used. To this end, 103 digital charts from 2018 were downloaded from https://www.inegi.org.mx/temas/topografia/default.html#. All vector layers were joined using the merge command. The continuity between water bodies represented in two charts or more was ensured employing a dissolve command. From the topographic information from INEGI, the cut lines that identify the water bodies were extracted individually.

Regarding the descriptive component of the spatial data, the identifiers and names from SIATL 3.2 were used. For the water bodies for which SIATL 3.2 did not have records, the identifiers and names of the INEGI vector data set were assigned. Those polygons that were categorized as PW from the G3WBM and that were not present in either of the two mentioned data sources were assigned a consecutive number.

2.3 Multiannual Area-Based Indicator Generation

By means of a spatial coincidence analysis procedure (outlined in Fig. 1) programmed in GRASS GIS 7.4.4 as a model, it was possible to obtain the area detected month by month for each water body, as well as the indicator of the area in percentage with respect to the surface stored in the multiannual RL. Essentially, with the GRASS GIS model, the binary water/no water mask of each month and the RL in raster format were processed. Through a spatial coincidence analysis between both spatial layers, the pixels that fall within the area of a particular water body registered in the multiannual RL were assigned their identifier and other descriptive characteristics such as the type of water body and its name. Cases were observed in which a water body in the RL appeared, for a particular month, divided into more than two conglomerates of pixels (individual water bodies). In these cases, the resulting polygons kept the identifier and the name of the water body in the RL to calculate its area with respect to this layer. For this, database management activities were carried out, summarizing, by means of SQL commands, the water bodies detected for each month that correspond to one in the multiannual RL. The management of the identifiers of these detected water bodies and the corresponding ones in the multiannual RL allowed the conjunction of both sources of information to obtain the indicators referenced to the multiannual RL.

The visual and statistical analysis of results was carried out in QGIS and R. As it is a large data set (big data), the results of the multiannual area-based indicator were synthesized and presented through maps and graphs on average for the 36 months of the study and for the seasons of the year applicable to the area according to Arreguín-Cortés et al. (2014).

3 Results

Figure 2 presents the RL obtained in the geographic context of the basins and their sub-basins. As can be seen, the connected water bodies and rivers are located in the lower part of the Basin.

3.1 Analysis of Cases of Permanent Water Bodies

Laguna Maluco and Laguna el Corcho are PW according to the RL. The detected extension of water bodies around them can be visualized for the four seasons of the three years of the study in Fig. 3. These seasons (Arreguín-Cortés et al. 2014) are winter rains (December to February), low water (March and April), tropical rains (May to August), and tropical winter rains (September to November). The water bodies are shown by colors: PW in blue, TW in pink, and WSV in orange. For comparison purposes, the area that each water body must occupy, according to its definition on the multiannual RL, is presented as a footprint, in the same colors previously mentioned for each type of water body, but with a diagonal lines pattern in the corresponding color.

As shown in Fig. 3, Laguna Maluco never totally covered the surface established in the multiannual RL. During 2016, the water body had a reduced coverage, and for the tropical and winter rains season of that year, it practically disappeared. Subsequently, it began an increase in its surface observed until the low water season in 2017, but in tropical rains of that year, it disappeared completely. During the tropical winter rains of that year, it recovered part of the previously lost flooded area. However, during all the seasons of 2018, its coverage was minimal and for at least three seasons of that year, it was fragmented into small water bodies. This case contrasts with that of Laguna El Corcho, which during the seasons of this study, showed coverage very close to the value stored in the multiannual RL.

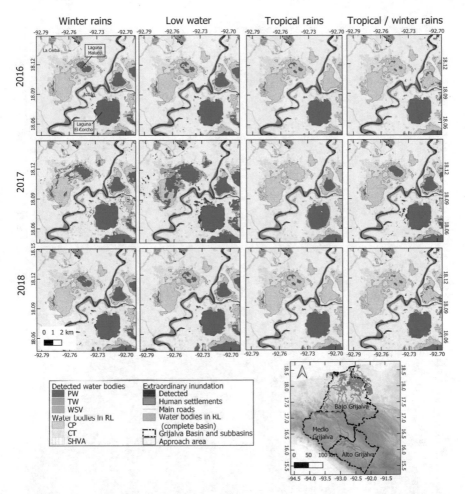

Fig. 3 Monitoring by the seasons defined in Arreguín-Cortés et al. (2014) from 2016 to 2018 of the changes in extension of two permanent water bodies (PW), located in the area with the highest concentration of water bodies in the lower basin. The Laguna Maluco shows many variations and El Corcho shows stability. TW: temporary bodies and WSV: wet soils and vegetation

3.2 Analysis of the Multiannual Area-Based Conservation Indicator

When viewing at the basin level, there appears to be little impact on water bodies. At this scale, larger water bodies such as great dams in the high and medium parts of the basin dominate the visualization and show little change. These large water bodies had on average for the 36 months, area values higher than 80% of the corresponding value in the multiannual RL (Fig. 4).

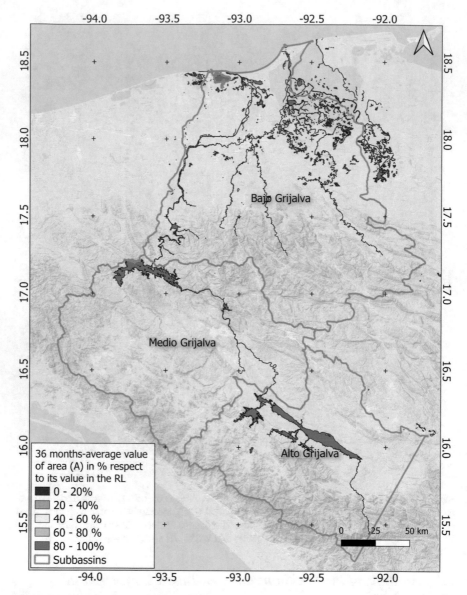

Fig. 4 36 months-average value of the flooded area (A) in % respect to its value in the multiannual reference layer (RL) of each water body visualized at the scale of the lower Grijalva basin

At an increased visualization scale, the situation about the area-based indicator for PW was different. The PW with low average values of the flooded area in % respect to its value in the multiannual RL (shown in Fig. 5 in red and orange) are located near human settlements and roads and on the edge of the Pantanos de Centla

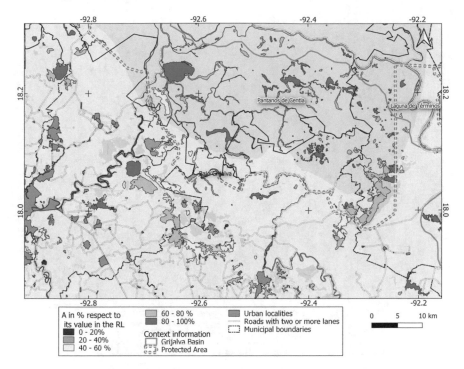

Fig. 5 36 months-average value of the flooded area (A) in % respect to its value in the reference multiannual layer (RL) of each Permanent water body (PW), visualized at the scale of the highest concentration of water bodies. This level of approach includes urban locations, roads, and municipal boundaries

protected area. In contrast, the best conserved PW (indicated in blue and green in Fig. 5) are located within the protected area.

At the scale of highest concentration of water bodies shown in Fig. 6, it can be observed that the TW are normally located outside or on the edge of the Pantanos de Centla protected area. They have reduced their flooded area the most as compared to the stored value in the multiannual RL. Even many of these (marked in red in Fig. 6) can be considered in risk to disappear.

At the basin level, the 36-months average value of the total area covered by water for PW is 1279 km^2. In the multiannual RL, this value is 1820 km^2. That means a reduction of 30% of the flooded area of PW. In the case of TW, this value is 45.3 km^2, which compared to its value in the RL (336 km^2) indicates a reduction of 86.5%. Regarding trends, it is observed that for PW the area sum tends to decrease slightly during the analyzed period. For TW and WSV this value tends to remain stable during the analyzed period. This can be seen in Fig. 7.

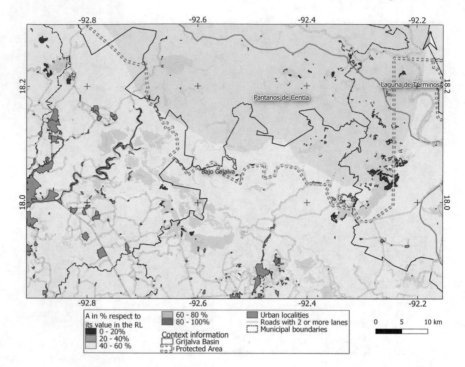

Fig. 6 36 months-average value of the flooded area (A) in % respect to its value in the reference multiannual layer (RL) of each Temporary water body (TW), visualized at the scale of the highest concentration of water bodies. This level of approach includes urban locations, roads, and municipal boundaries

3.3   Relevance of the Results in Decision Making and Public Policies

Information on the presence and conservation of water bodies is a fundamental element to support the spatially differentiated decision-making process of the water resource. To discuss the above, we return to the analysis of some of the figures of those previously presented.

As shown in Fig. 3, seasonal monitoring of changes in particular water bodies, especially PW, is a fundamental tool for the decision-making process in terms of knowing which are suffering damage and which are maintaining values close to the area stored in the multiannual RL.

Figures 5 and 6 show the location of human settlements and roads and water bodies visualized according to their value of the area in % of this value in the multiannual RL, for the zone with the greatest presence of CA located in the lower Grijalva basin. The water bodies with the lowest values of area in % of the corresponding value in the multiannual RL are those located in the vicinity of human settlements and roads. This area includes municipalities such as Centro, Reforma, and Nacajuca. Likewise,

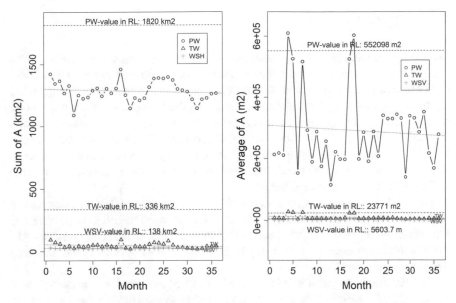

Fig. 7 Monthly values of sum (left) and average (right) of area (A) for all the water bodies in the basin, grouped by type. Black dotted lines indicate the values of these parameters in the multiannual reference layer (RL). PW: permanent water bodies, TW: temporary bodies, WSV: wet soils and vegetation. Blue dotted are the trend lines for each time series

the water bodies in this area are smaller than those observed in the area with the highest presence of CAs located in locations far from human settlements and roads.

These are the water Bodies observed within municipalities such as Centla and Macuspana and they are the ones with the highest proportions of area in % to this value in the multiannual RL. This condition can be related to the fact that the most affected area is where roads and human settlements are accessing and decreasing the surface of the water bodies. In these areas, the urbanization process is increasing as well as its harmful effects on the environment. In terms of public policies, the urbanization process should be controlled to promote the protection of the affected water bodies and the conservation of those with lower affectation.

In this context, the Pantanos de Centla protected area has served as a buffer zone for the growth of the urban area. Particularly, it has been a protection zone for water bodies, given that those located in its interior remain with areas close to the corresponding values in the multiannual RL. The design of public policies for the strengthening and maintenance of ANPs is recommended.

During the research, geospatial information was generated to support the design of spatially differentiated policies that allow the conservation of the water bodies and the restoration of those that can lose their persistence. This information is available to decision-makers and public policymakers.

4 Conclusions

During the research process, large amounts of data (big data) were processed. The applied remote sensing and spatial analysis procedures allowed to obtain a multiannual and area-based indicator of the conservation of each water body. Using an RL obtained in multi-year terms allowed quantifying the changes that each water body experienced on a multiannual basis and generating indicators about its conservation degree based on the area covered with water. It was possible to quantify the changes in terms of the area of the water bodies based on a long-term reference. However, it is necessary to confirm the results obtained by using another multiannual RL. This will be addressed in future research.

Regarding discoveries, a reduction of 86.5% of the area of TW was detected in multiannual terms. This can be a call for attention to avoid their loss of persistence. PW remain with values higher than 80% of the area stored in the multiannual RL. For the total area covered with water in the basin, a loss of 30% in the last three decades was obtained.

The Pantanos de Centla protected area has served as a barrier to preserve water bodies located inside it. However, the water bodies located around the limits of the protected area, which are close to towns and main roads, are being affected in terms of the observed condition of the area as compared to the same value in the multiannual RL. The developed methodology and the generated area-based multiannual conservation indicator make it possible to assess the state of water bodies at a regional level and identify particular water bodies that require urgent attention.

Acknowledgements The authors thank the Mexican research project "PROYECTO FORDECYT-2018-10. ANALYSIS AND MONITORING OF THE GROWTH OF THE URBAN ENVIRONMENT AND THE BEHAVIOR OF WATER BODIES FROM A SUSTAINABILITY APPROACH; CASE STUDIES: CORREDOR METROPOLITANO CENTRO PAÍS Y CUENCA DEL GRIJALVA", its coordinator Dr. Elvia Martínez, its collaborators and the financing fund CONACYT. This article is a product of that project.

References

Arreguín-Cortés FI, Rubio-Gutiérrez H, Domínguez-Mora R, Luna-Cruz FD (2014) Análisis de las inundaciones en la planicie tabasqueña en el periodo 1995–2010. Tecnología y ciencias del agua V(3):5–32. ISSN 2007-2422. http://www.scielo.org.mx/pdf/tca/v5n3/v5n3a1.pdf
Ellison AM (2004) Wetland of central America. Wetl Ecol Manag 12:3–55. ISSN 1572–9834. https://doi.org/10.1023/B:WETL.0000016809.95746.b1
López-Caloca AA, Tapia-Silva FO, López F, Pilar H, Ramírez González AO, Rivera G (2020) Analyzing short term spatial and temporal dynamics of water presence at a basin-scale in Mexico using SAR data. GIScience & Remote Sens 57(7):985–1004. ISSN 1548–1603. http://dx.doi.org/ https://doi.org/10.1080/15481603.2020.1840106
Olofsson P, Foody GM, Herold M, Stehman SV, Woodcock CE, Wulder MA (2014) Good practices for estimating area and assessing accuracy of land change. Remote Sens Environ 148:42–57. ISSN 0034-4257. https://doi.org/10.1016/j.rse.2014.02.015

Pontius Jr RG, Santacruz A (2014) Quantity, exchange, and shift components of difference in a square contingency table. Int J Remote Sens 35(21):7543–7554. ISSN 0143-1161. http://dx.doi.org/https://doi.org/10.1080/2150704X.2014.969814

Sánchez AJ, Salcedo MA, Florido R, Mendoza JD, Ruíz-Carrera V, Álvarez-Pliego N (2015) Ciclos de inundación y servicios ambientales en Grijalva-Usumacinta. ContactoS 97:5–14. http://www2.izt.uam.mx/newpage/contactos/revista/97/pdfs/inundacion.pdf

Yamazaki D, Trigg MA, Ikeshima D (2015) Development of a global ~90m water body map using multi-temporal Landsat images. Remote Sens Environ 171:337–351. ISSN 0034-4257. https://doi.org/10.1016/j.rse.2015.10.014

Zamora Saud N, Pérez Sánchez E, Carballo Cruz VR, Galindo Alcántara A (2018) Dinámica de las terrazas fluviales en la subcuenca Grijalva-Villahermosa, México. Boletín de la Sociedad Geológica Mexicana 71(3):805–817. ISSN 1405-3322. http://dx.doi.org/https://doi.org/10.18268/BSGM2019v71n3a10

Applications of Geospatial Data

Social Vulnerability Analysis of Three High Mountain Villages at Pico de Orizaba Volcano, Mexico, Using PCA

Angel de Jesús López-González, Katrin Sieron,
Karime González-Zuccolotto, and Sergio Francisco Juárez-Cerrillo

Abstract This paper presents the results of a study determining the social vulnerability to natural phenomena in the localities of Rincón de Atotonilco, Atotonilco and Mesa de Atotonilco, in Veracruz (Mexico), on the northeastern flanks of a potentially active stratovolcano. The total population of the three localities is 1,283 inhabit-ants and all three show a very high degree of social marginalization (INEGI 2010). The villages are located in the middle of the Jamapa River ravine, one of the major gorges draining the volcano. Hence, the three locations are constantly affected by hydrometeorological and geological phenomena. In knowledge of the natural affec-tations that occur in the mentioned localities and given the intrinsic characteristics of the population, this work presents the results of a principal com-ponents analysis applied to the set of variables obtained in the field through a survey that was applied to a sample of inhabitants; a method adapted after a previously published work. Data on both, qualitative and quantitative variables were collected, in order to later calculate a vulnerability index at the household level. The results include a primary characterization of the three localities through a descriptive analysis of the surveys and the secondly the levels of vulnerability were obtained. Based on this, factors that increase vulnerability in the three localities were identified, such as the meager access to remote telecommunications, the distribution of ranches (by heritage) that encourages the invasion of the riverbed and the lack of prevention plans; mitigation measures are also proposed, which are aimed at developing prevention plans. The methodology used, is easily replicable at different scales and can be adapted to the intrinsic characteristics of each study area.

A. de Jesús López-González (✉) · K. González-Zuccolotto
Centro de Investigación en Ciencias de Información Geoespacial, Contoy 137, Col. Lomas dePadierna, Alcaldía Tlalpan, 14240 Ciudad de México, México

K. González-Zuccolotto
e-mail: karime.gonzalez@centrogeo.edu.mx

K. Sieron · S. F. Juárez-Cerrillo
Universidad Veracruzana, Lomas del Estadio s/n, 91000 Xalapa, Veracruz, Mexico
e-mail: ksieron@uv.mx

S. F. Juárez-Cerrillo
e-mail: sejuarez@uv.mx

© The Author(s), under exclusive license to Springer Nature Switzerland AG 2022
R. Tapia-McClung et al. (eds.), *Advances in Geospatial Data Science*, Lecture Notes in Geoinformation and Cartography, https://doi.org/10.1007/978-3-030-98096-2_8

Keywords Vulnerability · Natural phenomena · Lahars · Pico de Orizaba
volcano · Principal Component Analysis (PCA)

1   Introduction

Natural phenomena inevitably occur; they can influence human life positively or
negatively, or not at all. In many cases, their occurrence is associated with disaster,
even though the natural phenomenon itself is only a material expression of nature.
Natural phenomena can be treated as hazards as long as there is an agent in a situation
of vulnerability; and we speak of a disaster when this natural phenomenon negatively
affects the population (Maskrey et al. 1993).

Since there are methodologies to quantify the hazard, there are also those that
at-tempt to quantify vulnerability, however many of these cannot easily be replicated
at different scales and in different territories. Some of the previous studies show
different perspectives to address vulnerability, and we find that especially the social
aspect is often not taken into account (e.g., Hernández et al. 2003; Reyes et al. 2005;
Lara et al. 2006; Bollin and Hidajat 2006).

Therefore, after revising several methodologies, for this study we replicated the
one developed by Bohórquez (2013), in which the author proposes the creation of
a social vulnerability index through an analysis of main components for the city of
Manzanillo, Colima. The greatest advantage of this methodology is the approach of
individual persons through surveys, allowing the characterization of this particular
study area at the flanks of Pico de Orizaba volcano. Useful, in general, is also the
quantification of characteristics as the vulnerability study could later be used for risk
determination (together with the preexisting hazard maps).

The localities of interest are located at an average altitude of 2,161 m above
sea level (masl) on the northeastern flanks of the volcano, the highest point in the
Mexican Republic and the third highest in North America at 5,636 masl. Pico de
Orizaba is an active volcano in repose; however it presents potential hazards not
only linked to volcanic eruptions, but also to the occurrence of secondary lahars and
mass wasting processes, which depending on the type (see Varnes 1978; Alcántara
Ayala 2001) are mainly triggered by rainfall (between 1500 and 1800 mm), seismic
activity, and/or steep slopes together with easily erodible geological units. The latter is
particularly important above the treelined where vegetation gets scarce (see Palacios
et al., 1999; Palacios and Vázquez-Selem 1996; Sieron et al. 2021; Chiarle et al.
2007) But especially within the Jamapa and other ravines, the occurrence of floods
and flows (lahars) is linked to extraordinary hydrometeorological events during the
rainy season between June and October that cause in-tense rains in short periods
of time. Recently, in 2003, intense rains on the south-ern flank of Pico de Orizaba
caused strong floods, which together with mud and sediments descended at high
within the Chiquito River sub-basin, causing numerous damage downslope when
passing through inhabited areas. This extraordinary flood-event (or diluted lahar)
also affected important oil infrastructure facilities, causing severe damage and even

fatalities. Later, in 2012, when Hurricane Ernes-to crossed Pico de Orizaba volcano, and caused a mass flow event on the upper Jamapa channel, which downslope again caused a lot of damage (Morales Martínez et al. 2016; Sieron et al. 2021). Other natural phenomena in the study area typically include hail and strong winds affecting houses and infrastructure in general.

Additionally, the fact that the volcano is at rest does not exempt it from the volcanic hazard potentially present (Sheridan, 2004).

2 Methodology

The original questionnaire was proposed by Bohórquez (2013), and here adapted and applied. The survey is organized into four sections: the first collects socioeconomic data from the sample unit, the second comprises information about land use, the third section collects data related to housing infrastructure, basic housing services and community services present in the locality, while the fourth and last section gathers data regarding threats or dangers recognized in the locality. In total the survey consists of 63 questions deriving in the same number of variables.

2.1 Data Collection, Management and Organization

For the present study, the household unit was chosen as the sample object. In order to recognize the precise number of homes in each of the localities at the current moment (2019), a remote individual identification was undertaken using Google Earth. Non-probabilistic random sampling using the "Random Points" GIS tool was carried out at different percentages to obtain a representative sample number, where the quantity and distribution of the dwellings in the three localities was considered. Final samples of 30% in Rincón de Atotonilco, 35% in Atotonilco and 25% At Mesa de Atotonilco were obtained.

Subsequently, the surveys were carried out, which were adjusted in the field depending on the availability of the respondent (choosing a neighbor house in case nobody answered). Then the data were organized and captured in spreadsheets for the posterior analysis with statistical software and GIS.

Index of social vulnerability to natural hazards (ISV) through the analysis of principal components (PCA).

The Principal Component Analysis (PCA) is a statistical technique for the synthesis of information or reduction of variables (e.g., Terrádez 2013) without losing in-formation. Thus the "principal components" will be a conjunction of the original variables. The original variables of this study are 63, which were reduced to 12 synthe-sized ones. With these 12 new variables, the PCA was carried out. The im-mediate result is the correlation matrix, which indicates how closely related the variables are to each other. Finally, an extraction matrix of main components is obtained. Four

Table 1 Ponderated values of variables (modified after Bohórquez 2013)

Key	Variable name	Ponderated value
DP	Depending population	0.615
PC	Grade of population concentration	0.438
AA	Antropic activity	0.789
INC	Income	0.989
ELVL	Education level	0.989
IHH	Informality of the houshold	0.980
ILC	Informality of the locality	0.980
KI	Knowledge of the individual	0.886
PI	Perception of the individual	0.398
HK	Historical knowledge	0.998
EPP	Existence of prevention plans	0.997
RC	Response of the community	0.997

main components were generated, LEPL (Level of Exposure by Population Location), LESC (Level of Exposure by Socioeconomic Conditions), LEET (Level of Exposure by Empathy with Threat) and LEIOT (Level of Expo-sure by Institutional Organization against the threat). For the construction of the Social Vulnerability Index to natural hazards (SVI), a partial indicator was extracted for each main component. Once these indicators were extracted for each main component, a linear combination was constructed, which resulted in:

$$\text{ISV} = 0.405 \, \text{LELP} + 0.610 \, \text{LESC} + 0.851 \, \text{LEET} + 0.108 \, \text{LEIOT}$$

Each of these new variables has weighted values (Table 1), which were originally proposed by Bohórquez (2013) and serve as exponents of the survey results. The variables were multiplied with the weighted values and a product of the previous results was made with the ISV to obtain the vulnerability for each location. With the latter, they were represented graphically in vulnerability maps. Using the In-verse Distance Weighted (IDW) GIS tool, density polygons were generated to weight areas according to the different levels of vulnerability presented (Table 3).

Table 2 Ponderated values PCA

LEPL = 0.615DP + 0.438PC + 0.789AA	LEET = 0.886KI + 0.398PI + 0.998HK
LESC = 0.989INC + 0.989ELVL + 0.980IHH + 0.980ILC	LEIOT = 0.997EPP + 0.997RC

3   Results

Results were obtained in two aspects, the first related to the descriptive analysis of the surveys and the second, the levels of vulnerability identified in the three localities.

In the three localities, the young and adult population predominates, however the number of children under 12 years of age is considerable. There is a dependent population (under 10 years old, over 75 years old, with some cases of disability and pregnancy) of 35%. There is a very low educational level in general, with the highest level of studies being primary school (total of ~ 6 years) in Rincón de Atotonilco and secondary school (another additional 3 years) in Atotonilco and Mesa de Atotonilco. The daily household income is very low, between 60 and 80 Mexican pesos; the National Council for the Evaluation of Social Development Policy (CONEVAL) establishes that a family of four is currently in a situation of income poverty if their monthly income is less than $ 11,290.80 (CONEVAL, 2017). In any case, the net income is variable; despite being considered daily, it is actually the income received three or four times a week, when they are hired to harvest for example. Regarding the land use, most of it is residential use and only 10% is of mixed type, meaning that the homes also function as small businesses, especially groceries. Regarding the type of housing, wooden walls with sheet roofs predominate; however masonry constructions with rigid roofs (sheet and masonry) were also identified during field work. Most of the houses do not have public deeds since they identify their lands as inherited and distribute it among the current members of the family. Hence, there is no correct delimitation of property. 94% of the houses have electricity and drinking water, although sometimes they do not have a constant supply. In the three locations there is a drainage system installed, however only 38% use it; the remaining 62% argue that the installation and connection of the drainage network from the home to the local system implies a monetary investment which they are not able/willing to make given their low income. Concerning the services in the localities, very few people have their own car, how-ever there is a system of taxis and collective vans to get to the nearest towns and cities (to get groceries, health care etc.). They do not have health centers in any of the analyzed villages; the closest is located in the town of Teopantitla, 5 min by car or 20 min on foot. The most common natural phenomena recognized by the inhabitants are river flooding, landslides, strong winds, and heavy rains. The level of historical knowledge and response to hazards of the inhabitants is generally high.

With the previously established vulnerability characteristics, the results regarding the vulnerability levels are as follows for each locality (Table 3)

3.1   *Rincón de Atotonilco*

In Rincón de Atotonilco, 48% of homes have a very high level and 52% have a high level of vulnerability. The area with the highest concentration of homes with very

Table 3 Characteristics of vulnerability levels

Vulnerability level: Very High
This level means that the dwellings have the most precarious conditions, their inhabitants have lower levels of education and low incomes, and there is a general ignorance of the potential threat of the identified natural phenomena

Vulnerability level: High
At a high level of vulnerability, most of the houses have sheet roofs and concrete walls, with education levels not higher than secondary school and with an incipient level of awareness of potential threats

Vulnerability level: Medium
In this case, the houses have a low level of informality, their inhabitants have better educational conditions and a better income. In addition they have knowledge and are aware of the potential threats

Vulnerability level: Low
At this level, the housing, education and income conditions are good, the inhabitants are aware of the potential threats, they have historical knowledge, and the authorities maintain a good relationship and coordination with the inhabitants of the locality

high vulnerability lies to the north of the village, with 6 of the total of 12. The houses with a high level of vulnerability (13 of 25) are mainly distributed around the main road, except for the group of four houses in the first square of the locality (northwest). The differences between one and the other are really small but significant; one very obvious is the fact that it is close to the town's main communication route, which gives it a more effective response capacity in any case of emergency.

3.2 Atotonilco

In Atotonilco, a division to the center-south is clearly observed where houses with very high vulnerability are concentrated (12 of 35, corresponding to 34%). It is an area where a high number of dwellings with resistant constructions are concentrated and the largest number of services in the locality are present. Nevertheless, it is also the area with the most antecedents in terms of natural phenomena and disasters in the recent past, which also increases the perception of hazard. This means, that if something catastrophic were to happen, that area would be practically evicted, hence the fear "of losing everything" is present within the population. All this is reflected with the highest vulnerability index in this area. In the center-north there are two houses with very high vulnerability as well, which is due to the steep slope behind these houses. To the east of Atotonilco we find only one home with a very high vulnerability index, the rest (23 homes) correspond to high vulnerability. Likewise, in the western area of Atotonilco, two isolated points of very high vulnerability are observed, mainly due to structural characteristics of the houses and steep slopes.

3.3 Mesa de Atotonilco

In Mesa de Atotonilco we found 36% of homes with very high vulnerability; which are distributed: 10% to the east, to the center the largest group including 26%. The area of high vulnerability predominates, concentrated in the center and west of the town with 56% of homes. And finally, 8% of houses that present a medium level of vulnerability, these being the only ones of the three localities; the houses present education- and economic levels above the average and more resistant construction materials.

In total, 38% homes were counted with a very high degree of vulnerability, 59% of homes with a high degree of vulnerability, 3% of homes with a medium level of vulnerability and no homes with a low level of vulnerability in the three villages. The IDW reflects a distribution of 27% of very highly vulnerable area, 72% of highly vulnerable area and 1% of moderately vulnerable area (Fig. 1).

3.4 Factors Increasing Vulnerability Levels

Certain trends were observed that increase the degrees of vulnerability in the localities studied. One of them is the limitation of remote communication, since only 16 homes have landlines and two have mobile phones. Telephony is the only effective means of communication to nearby towns. On the other hand, the change of land use; from agricultural to residential use by inheritance and distribution of ranches encourages the invasion of the riverbed. Another factor is the increase of infrastructure in the central area of Atotonilco caused by local tourism. Another important and very determining factor identified during this study was the apparently not effective governmental management and detachment of the pertinent authorities with the localities; there are no prevention plans (known) in any of the localities. Another factor is the repetitiveness of the disaster; the surveys showed that the residents have a high empathy with the natural phenomena of the region, which means that they know them and know how to face them from past experiences. However, there is a peculiarity, despite the fact that disasters destroy their houses, they rebuild on the same site or contiguously, mainly because they have nowhere else to go.

3.5 Proposals of Vulnerability Mitigation Measures

With the information obtained during the surveys and analyses, it was stated that there are variables that determine to a greater extent the high levels of vulnerability of the dwellings in question, such is the case of the apparent non-existence of prevention plans. For this reason, we decided to create different scenarios where the influence of this variable on the levels of vulnerability is corroborated. The value of the variable of

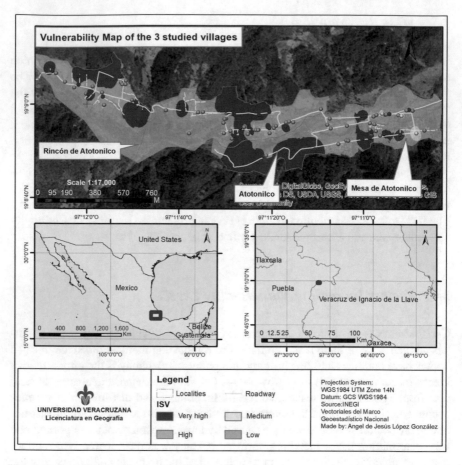

Fig. 1 Vulnerability Map of the 3 studied villages

the prevention plans was 1 "non-existent" in the three localities (of the three possible ones: 1 "non-existent", 2 "incipient", 3 "sufficient"). If the value were 2, "incipient", in the three localities, the levels of vulnerability would be lower, for example, in Rincón de Atotonilco instead of having 48% of houses with very high vulnerability and 52% with high vulnerability, this would be reduced to 32% of homes with very high vulnerability, 60% with high vulnerability and 8% with medium vulnerability. In the case of Atotonilco, instead of showing 34% of homes with very high vulnerability and 66% with high vulnerability, the situation would change to 29% of homes with very high vulnerability and 71% with high vulnerability. In Mesa de Atotonilco, instead of having 36% of homes with very high vulnerability, 56% with high vulnerability and 8% with medium vulnerability, there would be 26% of homes with very high vulnerability, 62% with high vulnerability and 13% with medium vulnerability. If the value were 3, "sufficient", in the three localities we would have a considerable decrease in vulnerability with respect to the results obtained. For example, in

Rincón de Atotonilco there would be 24% of houses with very high vulnerability, 60% with high vulnerability and 16% with medium vulnerability, in Atotonilco there would be 23% of homes with very high vulnerability, 66% with high vulnerability and 11% with medium vulnerability, in Mesa de Atotonilco there would be 20% of homes with very high vulnerability, 64% with high vulnerability, 13% with medium vulnerability and 3% with low vulnerability (Table 4). Given the above hypothetical results, the creation of prevention plans in the localities is considered pertinent and for these plans to be satisfactorily fulfilled, a good government intervention is required, and therefore it is recommended to seek a greater connection of the relevant authorities. (e.g., Civil Protection) and the localities in question, something that until now is non-existent according to the words of the inhabitants themselves. This same government intervention would promote the correct equipment of the localities to reduce informality, and hence also supplying the other factors of in-creased vulnerability.

4   Conclusions

Studies referring to vulnerability are a subject of multiple perspectives, which generates that they are approached from different currents of research (social, environmental, political, etc.); however they lack comprehensiveness, that is, many of them focus the attention of the method on a single thematic line, which generates complication when trying to replicate methodologies at different scales and areas of study. With the methodology of Bohórquez (2013), both social and physical aspects were taken into account, also resorted to the political and organizational sphere which allowed, through the PCA, to characterize and measure the vulnerability in a complicated study area. The survey tool was used, which allowed us to have a more natural approach with the population and thanks to this it was possible to punctually characterize and understand the particularities of the study area. Statistical analysis was applied, using quantitative and qualitative modeling, which allowed us to integrate a large number of variables demonstrating that the vulnerability assessment is not limited to a single approach and that each variable contributes in very specific measure to the construction of vulnerability. The main limitation found in the application of this methodology is that, although the generation of a database gives us the advantage of reducing the dimensionality to a more representative sample unit, this creation of our own database, it depends to a great extent on the quality of the data collected through the surveys, which are not always openly available, which could generate the need to adjust the methodology according to this availability. In conclusion, vulnerability levels are generally very high and high in the studied villages, which would be the first to be hit, when flooding or lahars occur in the Jamapa gorge. Apart from education, income and structural issues, particularly problematic is the inheritance of terrain. Splitting the land in several pieces depending on the number of children, leads to subsequent invasion of the river influence area with time.

Table 4 Comparison between prevention plans values

Prevention plans values	Non-existent	Incipient	Sufficient							
Vulnerability levels	Medium	High	Very high	Medium	High	Very high	Low	Medium	High	Very high
Rincón de Atotonilco	0	13	12	2	15	8	0	4	15	6
Atotonilco	0	23	12	0	25	10	0	4	23	8
Mesa de Atotonilco	3	22	14	5	24	10	1	5	25	8

Further, the impact of specific low valued variables (as prevention plans) has been pointed out. Vulnerability values established here, could be potentially used for risk analysis, using preexisting hazard maps (for lahars for example).

Acknowledgements Funding was provided partly by Geoscientists Without Borders® (GWB) Program of the Society of Exploration Geophysicists (SEG), through the project "Hydrometeorologic and geologic hazards at Pico de Orizaba volcano, Mexico".

References

Alcántara Ayala I, Echavarría Luna A, Gutiérrez Martínez C, Do-mínguez Morales L, Noriega Rioja I (2001) Inestabilidad de laderas. In: Ines-tabilidad de laderas (pp 36–36)

Bohórquez JET (2013) Evaluación de la vulnerabilidad social ante amenazas naturales en Manzanillo (Colima). Un aporte de método. Investigacio-nes Geográficas 81(Mx):79–93

Bollin C, Hidajat R (2006) Community-based risk in- dex: pilot im-plementation in Indonesia. In: Birkmann J (ed) Measuring Vulnerability to natural hazards. United Nations University Press, Tokio, Japon, Towards disasters resilient societies, pp 271–289

Chiarle M, Iannotto S, Martara G, Deline P (2007) Recent debris flow occurrences associated with glaciers in the Alps. Global Planet Change 56:123–136

Hernández L, Cruz H, Márquez B, Suarez C, Padlog M, Palomar P (2003) Aproximación al análisis de la vulnerabilidad del Volcán de Colima (Ja-lisco, Mexico), Vegueta, México, pp 241–254

Instituto Nacional de Estadística e Informática (INEGI) (2010) Prontua-rio de Información Geográfica Municipal de Calcahualco, Veracruz, México

Lara L, Clavijero J, Hinojosa M, Huerta S, Wall R, Moreno H (2006) NVEWS-Chile: Sistema de clasificación semi-cualitativa de la vulnerabili-dad volcánica. Universidad Tecnológica Metropolitana, Chile

Maskrey A, Cardona O, García V, Lavell A, Macías JM, Romero G, Chaux GW (1993) Los desastres no son naturales

Morales Martínez MA, Welsh Rodríguez CM, Ruelas Monjardín LC, Weissling B, Sieron K, Ochoa Martínez CA (2016) Afectaciones por posible asociación de eventos hidrometeorológicos y geológicos en los municipios de Calcahualco y Coscomatepec, Veracruz. Teoría y Praxis, 12(Especial, octubre 2016), pp 31–49

Palacios D, Vázquez-Selem L (1996) Geomorphic effects of the retreat of Jamapa glacier. Pico De Orizaba Volcano (mexico) 78(1):19–34

Reyes C, Flores L, Pacheco M, López O, Valerio L, Zepeda R (2005) Evaluación de la Vulnerabil-idad. Capítulo VIII. Evaluación simplificada de la vulnerabilidad de la vivienda unifamiliar ante sismo y viento. Centro Nacional para la Prevención de Desastres, México, pp 311–338

Sieron K, Weissling BP, Morales-Martínez MA, Teran S (2021) Reconstruction of the upper slope conditions of an extraordinary hydro-meteorological event along the Jamapa glacier drainage system Citlaltépetl (Pico de Orizaba) Volcano Mexico. Front Earth Sci 9. https://doi.org/10. 3389/feart.2021.668266

Detection of *Phoradendron Velutinum* Implementing Genetic Programming in Multispectral Aerial Images in Mexico City

Paola Andrea Mejia-Zuluaga, Leon Felipe Dozal-García, and Juan Carlos Valdiviezo-Navarro

Abstract This research implements Genetic Programming to design a spectral index that allows the automated detection of the species *Phoradendron Velutinum* because it is a pest that leads to the detriment of forest health causing serious damage to the host trees. Employing multispectral aerial images taken in the field, pre-processed and selected for the creation of a set of masks with the presence of the pest, together with the use of different terminals and functions, it was possible to obtain an algorithm capable of classifying mistletoe with 96% overall accuracy and a fitness value (Weighted Cohen's Kappa = 0.45) on the test data set. Additionally, a comparison was made with the Structure Intensive Pigment Index 2—SIPI2 for the detection of *P. velutinum*, the results show that SIPI2 does not allow the correct identification of this particular pest.

1 Introduction

Phoradendron velutinum, also known as true mistletoe, is a hemiparasitic shrub plant of aerial parts of trees and shrubs, which has a preference for hosts in broadleaf forests and sometimes with mixed conifers. The spread of this pest has become a problem because it uses a ballistic propulsion method and some animals transport the seed to new host trees, making it difficult to detect new shoots.

P. A. Mejia-Zuluaga (✉) · L. F. Dozal-García · J. C. Valdiviezo-Navarro
Geospatial Information Science Research Center - CentroGeo, Mexico City, Mexico
e-mail: pmejia@centrogeo.edu.mx

L. F. Dozal-García
e-mail: ldozal@centrogeo.edu.mx

J. C. Valdiviezo-Navarro
e-mail: jvaldiviezo@centrogeo.edu.mx

One of the regions affected by this pest is the conservation soil associated with the *San Bartolo Ameyalco* Community located south of Mexico City, where, the uncontrollable presence of the true mistletoe has generated an environmental problem because by reducing the vitality of the trees, and ecological imbalance is presented which leads to a decrease in carbon sequestration and puts the balance of aquifer recharge at risk. and springs that depend on this forest area. Consequently, this region is socially and economically impacted because it is the conservation land that contributes the greatest amount of water to the Valley of Mexico, in addition to reducing the contribution to the Environmental Services administered by the community.

Therefore, this research pursues the objective of designing a spectral index for the automated detection of *Phoradendron velutinum*, generated from the genetic programming and multispectral images, to facilitate early detection and promote adequate control.

Genetic Programming—GP as a tool inspired by evolutionary algorithms, developed to automate tasks and solve different user-defined problems, has been implemented in areas such as regression analysis (Chen et al. 2005), prediction (Huan-rong et al. 2010; Huo et al. 2007), classification (Brameier and Banzhaf 2001), symbolic regression (Icke and Bongard 2013), optimization (Yuan et al. 2008), among others. The diversity of applications is possible since the algorithms of GP perform an abstraction of the knowledge represented in mathematical expressions or systems based on rules, in a probabilistic search space inspired by the Darwinian theory of evolution.

Contributions

Among the contributions made in this research, there is a new approach to the study, detections, and classification of forest pests through evolutionary techniques such as Genetic Programming.

On the other hand, it provides a classification spectral index sufficiently precise to detect the presence of the pest *Phoradendron Velutinum* in the Conservation Soil of Mexico City, and given the characteristics and information requirements for the application of this index, it is possible to do it with conventional aerial RGB images, which implies a reduction in costs and acquisition of equipment for forest sanitation entities such as the Natural Resources and Rural Development Commission—CORENA, with whom We had the opportunity to limit and address the study problem due to mistletoe infestation in the region.

Related Work

There are various methodologies to study pest detection, where the use of aerial images obtained from drones have allowed the inclusion of various supervised learning techniques (Lee et al. 2019; Dwivedi et al. 2021; Shankar et al. 2018).

In the case of forest pests, there are studies on the detection of the bark beetle in mixed forests (Minařík et al. 2020) and the spectral detection of *Dendrolimus tabulaeformis* (Zhang et al. 2020). Additionally, there is research regarding forest parasitic plants such as mistletoe, such is the case of the work reported by Sabrina et al. (2020) in which multispectral aerial images were used with a Convolutional

Neural Network; in this work, they place special emphasis on the architecture of the ANN as an approach to these pests, in the same way, the detection of these species with the use of hyperspectral images was done through the classification of the Spectral Angle Mapper—SAM with a value threshold of 5° considering the spectral signature of the parasitic species. On the other hand, Pernar et al. (2007) reported through a study of mistletoe in coniferous forests in Croatia, that it is possible to detect mistletoe species through a near-infrared digital camera and a combination of supervised and unsupervised learning techniques.

In Mexico, a study was carried out (León Bañuelos 2019) for the detection of the dwarf mistletoe *Arceuthobium globosum* present in the Nevado de Toluca, exploring the implementation of RGB images taken by drone, the supervised classification using the algorithm KNN and colorimetry algorithms. This research is relevant because, in addition to verifying the viability of supervised learning in the exploration and detection of parasitic plants, it also determines under a study that for the detections of forest pests in aerial images, it is necessary to have a spatial resolution less than or equal to 10 cm/px, given the physiological composition of these plants and the little contrast that it could present concerning the host trees.

2 Materials

2.1 Study Area

The study area is located in the conservation land south of Mexico City, registered as Community Conservation Area—ACC of the San Bartolo Ameyalco Community located at coordinates 19°20'N and 99°16'W at a height of 2, 420 *masl* in Álvaro Obregón's Delegation (Fig. 2). That ACC was registered in the Community Areas for Ecological Conservation program in 2017, to protect, improve, and conserve the natural resources and environmental services that this area provides to Mexico City.

The area of the ACC in San Bartolo Ameyalco is made up of 244 hectares, most of which are cataloged as forest area, due to the presence of oyamel *(Abies religious)* forests in 23% of the surface and pine *(Pinus spp)* at 38%. There is also the presence of forests of cedar *(Cupressus)* and oak *(Quercus spp)*, on the other hand, 22% of the surface is used for rainfed agriculture.

2.2 Phoradendron Velutinum

Phoradendron Velutinum (DC.) Oliv.—*P. Velutinum* is a forest pest also known as Barbas, bungu, graft, evil eye, mistletoe, or on the stick (Universidad Nacional Autónoma de México 2022), which is part of the *true mistletoe* species that are normally found in temperate zones of Mexico City (Gonzáles Gaona et al. 2017).

Fig. 1 **a** red circle indicates the presence of *P. Velutinum* at different levels of the host tree. **b** blue circle shows the establishment of the haustorium

P. velutinum is a hemiparasitic shrub with branched stems, opposite and decussate leaves, inflorescences in the form of articulated pedunculated spikes. It generally occurs in the aerial parts of trees or shrubs because the location in the higher parts allows it to obtain sufficient illumination for the photosynthesis process (Fig. 1), and due to its hemiparasitic condition, it is capable of extracting compounds from their host and photosynthesizes organic matter (Gutiérrez Vilchis and Reséndiz-Martínez 1994).

Physiologically, it is up to 80 cm long, internodes 8 cm long, without cataphiles, and with a petiole of 5–20 mm long, the coloration of this parasite is green or yellowish with a velutine pubescence frequently yellow in the young parts and the mature parts glabrescent. It has male inflorescences of 2.5 cm long, 2 to 5 segments of 6 to 40 flowers each in 6 longitudinal rows, the female inflorescence is 4 cm long with fruit, 2 to 4 segments of 12 to 30 flowers per segment in 6 rows (Rzedowski and de Rzedowski 2011).

This species has an establishment stage, in which the seed with chlorophyll endosperm produces simple sugars as a source of energy before germination. Afterward, the incubation stage follows, in which a radicle develops that penetrates the host cortex until it reaches the vascular tissues where it develops cortical haustoria (0.8–12 cm) (Mathiasen et al. 2008). Once established with aerial stems and flowers, a pollination process begins that can take 4 to 6 weeks. Finally, it disperses the seed through ballistic propulsion, that is, projecting the mature fruit around 15 m, which allows said seed to adhere to the branches of a new host; other means of dispersal are through the feces of birds and squirrels that feed on mistletoe (Alvarado-Rosales et al. 2007).

Table 1 P4 Multispectral features

Features	
Max Speed	50 km/h (31 mph) (P-mode)
	58 km/h (36 mph) (A-mode)
Max Flight Time	Approx. 27 minutes.
GSD*	(Aircraft altitude/18.9) cm/pixel
GNSS	GPS + BeiDou + Galileo
Camera	Six 1/2.9" CMOS
	Max Image Size: 1600×1300 (4:3.25)
	Photo Format: JPEG (visible light imaging) + TIFF (multispectral imaging)
Lens	Focal Length: 5.74 mm (35 mm format equivalent: 40 mm), autofocus set at ∞

*Ground Sample Distance (GSD)

2.3 Data Collection

2.3.1 Surface Sampling

The data collection in the field was carried out on April 13, 2021, at 12:30 p.m. with the support of the Commission on Natural Resources and Rural Development—CORENA technical team, who have information on the location of trees infested by *P. Velutinum* in the area.

At a solar elevation of 57.58° and an azimuth of 103.45°, aerial multispectral images of the affected regions were acquired using a P4 Multispectral Unmanned Aerial Vehicle—UAV, used in precision agriculture applications, environmental monitoring, and in-plant inspection and maintenance. This equipment has six 1/2.9" CMOS sensor camera and a 2 MP global shutter with stabilizer on three axes, composed of an RGB sensor (R (450 ± 16nm) G (56 ± 16nm) B (650 ± 16nm)) for the visible spectrum range and five monochrome sensors that include coverage in the Red Edge spectral band—REG (730 ± 16nm) and Near-Infrared—NIR (840 ± 26nm). Table 1 presents the general characteristics of the UAV.

To cover the largest surface in the shortest flight time, the study area was divided into four polygons (see Fig. 2 and Table 2), the altitude range was kept between 80 and 110 m in height to acquire a GSD value lower than 10 cm, which is the spatial resolution suggested by (León-Bañuelos et al. 2020) for the detection of pests at the tree level. Figure 2 shows the distribution of the flight polygons, which cover a total area of 147.25 ha, which were recorded through 2,565 images each made up of 3 individual files in the bands (RGB, REG, and NIR). The full estimated flight time was 1 hour with 40 seconds.

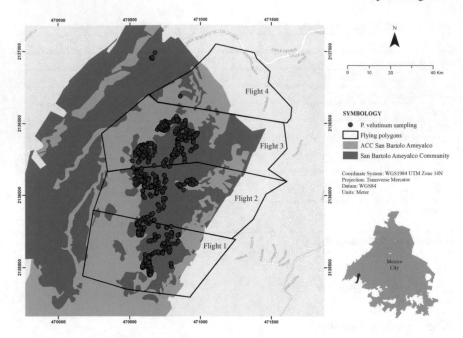

Fig. 2 Flight polygon distribution and sampling of *P. Velutinum*

Table 2 Data collection Design—UAV

Polygon	Area (ha)	Flying height (m)	GSD (cm/px)	Number of images
1	31.40	110	5.8	474
2	46.34	80	4.2	1,244
3	49.14	90	4.8	522
4	20.37	110	5.8	325

2.3.2 Multispectral images

A multispectral image I can be defined as a data cube or pixels $I(x, y, \lambda)$ where x and y represent the coordinates in the plane, column, and row, respectively, and λ represents the depth, that is, the spectral band. In the case of multispectral images, the bands can reach a maximum of 20 (Thigpen and Shah 2008). These images can be obtained by means of filters or sensors sensitive to specific wavelengths, in order to capture different ranges of the electromagnetic spectrum. Usually, ranges of the visible spectrum, are recorded by the combination of the red (R), green (G), blue (B), and infrared bands.

The pixel $I(x1y1, \lambda)$ contains an intensity value called Digital Number (ND) translated from the radiance captured by the sensor (Thigpen and Shah 2008). The ND response can be recorded at different wavelengths that correspond to different bands, or ranges, into which the Electromagnetic Spectrum is divided cite Emery2017, and depends mainly on the chemical and physical composition of the observed object (Ron Schmitt 2002).

In the case of multispectral images captured with cameras mounted on UAVs such as the *Sequoia* camera, *MicaSense*, or the *P4 Multispectral* sensors, they record the NDs with a spectral resolution of 16 bits (216) which allows storing a greater variety in the grayscale.

2.3.3 Vegetation Index

Given the importance of studying the earth using Remote Sensing techniques, in recent decades the use and applications of spectral information recorded from different sensors such as LiDar, spectroscopes, multispectral cameras, among others, have been deepened. This has led to the creation of spectral indices in the area of Remote Sensing, which, based on mathematical operations, establish relationships between different spectral bands, through which it is possible to detect specific surfaces and patterns. For example, there is a wide range of vegetation spectral indices, which in recent years have been used to detect changes in land cover (Xue et al. 2021), growth models of crops such as cotton (Yeom et al. 2017), or the estimation of the leaf area index (Xie et al. 2014), just to mention a few applications.

The Structure Intensive Pigment Index 2—SIPI2 (Peñuelas and Filella 1995) is one of the indices focused on forest analysis with high foliage variation, this index is related to the structural intensity of the plant's pigmentation (Mihaylov et al. 2020). The computing of this index is carried out using specific wavelengths (800nm, 505nm, 690nm) so it is generally used in satellite images, however, for multispectral images captured with UAV, the implementation is carried out employing the Eq. 1 (Qi et al. 2021; Peres and Cancelliere 2021).

$$SIPI2 = \frac{NIR - Green}{NIR - Red} , \tag{1}$$

2.3.4 Computer Equipment

The stage of preprocessing the images to create the data set with the presence of *P. velutinum* (see Sect. 3.2), was performed on a WorkStation Intel (R) Xeon (R) W-2102 CPU @ 2.90GHz, 16 RAM and Windows 10 operating system. The implementation of GP to find an optimal solution to *P detection. velutinum* in aerial images (see Sect. 3.3), was carried out in Software MATLAB R2019a implementing GP-Lab cite Silva2007, on a server Intel (R) Xeon (R) CPU. E5-2690 v2 @ 3.00GHz 250 RAM and Ubuntu 18.04.5 LTS operating system (GNU / Linux 4.15.0-122-generic x86-64.

3   Methods

3.1   Genetic Programming

Genetic Programming—GP is a supervised learning technique established in 1992 by John Koza (Koza 1992) that aims to generate computer programs automatically, is part of Evolutionary Computing, which is a subfield of Computational Intelligence, a branch of Machine Learning and Artificial Intelligence. In Evolutionary Computting, the process of evolution and natural selection described by Charles Darwin in 1859 (Darwin 1859) is described by the survival of the fittest individual where its characteristics are transmitted to the next generation through reproduction and Genetic inheritance is used as a model and as a strategy to find optimal or close to optimal solutions to a given problem. Evolutionary techniques are generally applied to optimization problems. Many of those problems are combinatorial optimization problems, which are computationally difficult (NP-hard).

A GP algorithm works as follows. Initially, a set of possible solutions or candidate solutions called *individuals* is randomly generated, which are evaluated based on an aptitude function that will show how good, or bad, each of these individuals is at the time of solving the problem. Once the aptitude of each individual is evaluated, is used to establish the probability of selection of reproduction of the individuals, that is, of being *Parents* of the next generation. The fitter the individual, the greater their probability of reproduction, this allows the best characteristics to be inherited to the following generations. In this process, the new populations will no longer be random but will be generated through a reproduction process using different *genetic operators* such as crossing and mutation, resulting in new individuals that will also be iteratively evaluated through several generations. until the best possible solutions are found (Koza 1994). Figure 3 shows the general workflow of the GP process.

Representation. An important characteristic of GP algorithms is that they are represented by a tree structure, called *genotype*, which allows increasing computational efficiency compared to their predecessors, Genetic Algorithms (Cramer 1985), since they are determined recursively. Within the tree structure, the root node and the internal nodes contain the functions assigned to solve the problem (operator), and the leaf nodes refer to the data that will be calculated (operand) (Bi et al. 2021), for this example, we will are represented by the spectral bands of the images.

Terminales and Functions. The *terminals* refer to the data or values that will be processed i.e. image bands, numerical variables, vectors, etc., while *functions* refer to the processes or operations that are performed on the terminals i.e. $\{+, -, *, /, cos, sin\}$, etc. Both the terminals and the functions are used to form the structure of the candidate solutions (trees) and must be oriented to establish a pertinent search space to develop the assigned task.

Initial Population. The initialization is carried out by creating the first population of individuals choosing their terminals and functions randomly. There are several procedures for generating the initial population, it is called a generative method, which can take the following directions: total initialization—*fullinit*, increasing

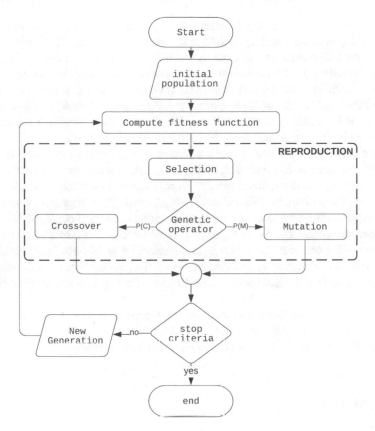

Fig. 3 GP workflow

initialization—*growinit* and ramping initialization—*rampedinit* (Silva and Almeida 2007).

Fitness Function. The calculation of the fitness value of each individual is fundamental for the evolution of the GP, because the "natural" selection, reproduction, and subsequent evolution through the generations, is based on inheriting the best characteristics that lead to perform in an optimal specific task In each generation, the performance of the n individuals of the population must be evaluated to solve the problems, the fitness function must give a certain degree of certainty regarding the problems (Morales and Casas 2007).

Selection. Once the fitness evaluation of each individual has been carried out, it is necessary to select the solutions for their reproduction, where the individuals with the best performance, have a greater probability of reproducing and passing their genetic material to the next generation. However, it is not enough to select the best solutions since when the characteristics of the *Parents* are inherited, the offspring tend to be similar to each other (Poli et al. 2008), which causes a loss of diversity that can be seen represented in a premature convergence.

In the process of selecting individuals, two important genetic parameters are considered (Chaudhary and Iqbal 2009): *sampling* and *elitism*. The sample allows to select the individuals who will act as parents of the next generation, this process can be done through the *roulette* methods, *tournament'*, *lexictour* or *sus*. On the other hand, elitism is how individuals are chosen for the next generation taking into account their evaluation of fitness, among the options is *halfelitism*, *totalelitism*, *keepbest* and *replace*. The technical specifications of each sampling method and elitism can be seen in greater detail in Silva and Almeida (2007).

Genetic Operators. Genetic operators are processes for generating new *children*, from existing *parents*, by combining their trees or substituting branches for new ones. In this study, we use two genetic operators, crossover and mutation. The selection of these genetic operators is assigned employing probabilities, trying to conserve the best genetic material while diversifying the solutions (Banzhaf et al. 1997).

Crossover: consists of the crossing of genetic information provided by two *Parents* to generate one or two *children* from the selection of subtrees (portions of trees assigned from a random node) in each individual, the selected subtrees are exchanged between one individual and another, resulting in offspring with new genetic characteristics.

Mutation: this process aims to introduce new genetic material to reduce bias and increase diversity, through the random selection of a node that will be replaced by new functions and randomly selected terminals (*Grow* method).

3.2 Pre-processing

3.2.1 Image Registration

Before training the GP algorithm, it was necessary to preprocess the images, starting with the registration of the RGB bands with the REG and NIR bands, since a mismatch was experienced between the exposure time of each sensor, causing spatial displacements between the bands of each photographic capture. This step is important for accurate spectral measurement of the presence of *P. velutinum*.

The bands were registered using the algorithm *Speeded-Up Robust Features*— SURF (Bay et al. 2006), this algorithm is designed as a detector and descriptor of local characteristics. It is based on *Scale-invariant feature transform*—SIFT (Lowe 1157), however, it presents improvements regarding the decrease in computational time and more robust support against the transformations on the image. This is because it uses box filters (using integral images (Crow 1984) to create different scale-spaces instead of iteratively reducing the size of the image as in other methods.

The main workflow of SURF consists of interest point detection, local neighborhood description, and point matching.

To carry out the correct registration between bands of the 2, 565 images taken with P4 Multispectral, and since SURF functions are composed of multiple parameters, it was necessary to perform an iteration of these criteria in a sample set of 100

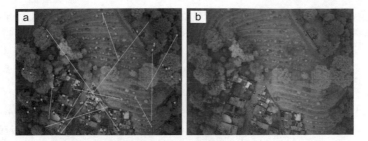

Fig. 4 a Points of interest between bands detected from SURF algorithm. **b** Registered Image Result

RGB images together with a random selection of their pairs in both the REG and NIR bands. The computational time to achieve the iteration of 14, 400 combinations was 21 days and 8 hours. The objective of this process was to find the parameters that maximize the registration between the RGB, REG, and NIR bands in a forest environment with a few buildings and roads. On the other hand, the metric used to evaluate the registry was the Structural Similarity Index—SSIM, this index evaluates the similarity between two images without compressing or distorting the reference image; the basis for this metric is a weighted comparison of luminance, contrast, and image structure.

The Fig. 4a, shows an example of the detection of points of interest detected by SURF where you can see the offset between the bands, in part b you can see the results of the successful registration of the images.

3.2.2   Mask Image Creation

Field trips were carried out around various sectors in the study area, to carry out visual training in the detection of *P. Velutinum*, both in its coloration and in its appearance and arrangement on the trees. The knowledge of these aspects in a physical environment as in a digital environment (photographs), is essential for the subsequent identification of the plague on the images.

The identification of the true mistletoe was carried out on the images through image interpretation, based on 885 sampling points GPS captured in field trips carried out by CORENA during the first months of the year 2021. This process resulted in the previous selection of a total of 94 images in which there is a presence of the pest in different proportions. In this selection, the reduction of overlap between exposures was considered to have unique data within the segmentation of the images.

This identification is accompanied by the creation of binary masks using ROI polygons, in which the zones with a value equal to 1 represent the presence of *P. velutinum* and the values equal to 0 represent the absence of it. This creation of masks was carried out through the application *Image Segmenter* (MATLAB 2019) implemented in the Image Processing and Computer Vision toolbox of MATLAB.

Through this application, it was possible to manually locate the sources of mistletoe infestation and segment the polygons present in the image.

3.2.3   Data Set Creation

Once the masks of the 94 images had been made, a division of the original data with a size of 1300 × 160 px was carried out in small pieces of 300 × 300 px. This process is intended to increase the volume and diversity of the data set with different perspectives of the pest on the images and to reduce processing time through the use of smaller data.

Next, it was necessary to make a selection of the samples that first had the presence of mistletoe, since the cut of each image was made in different views sequentially and proportionally to it, many of the samples only refer to the background of the territories (understood as any surface that is not *P. velutinum*). The second requirement implemented in this selection was the balance of classes (Goncalves and Silva 2013), since in some sections the polygon that indicated the presence of the pest represented less than 10% of the total image. manually, considering the details and context of each image.

It is important to highlight that the creation of this data set is based on original information taken in the field to address the specific study of the parasitic plant *Phoradendron velutinum* present in the study area and on the identification, segmentation, and selection by part of the expertise acquired in this work.

3.2.4   Threshold Selection SIPI2

Since Structure Intensive Pigment Index 2 is not a specific index for the classification of mistletoe but rather they present a wide-scale thought in the variety of foliage, it is not possible to determine *a priori* which is the proportion of the index that could best determine the presence of *P. velutinum*. Therefore, a series of threshold tests were carried out to determine which is the most appropriate range for the detection of this pest. This thresholding was performed on 100 randomly selected images within the data set designed for mistletoe (multispectral tiles and their respective mask), the iteration contemplated 10 threshold values ranging from 0 to 0.9.

As a result, it was obtained that the threshold that could best classify the presence of mistletoe is 0.9, a value of $k_w = 0.0047$, $k = 0.0059$, and an $OA = 0.5383$ (Fig. 5).

3.3   GP *Implementation*

The implementation characteristics used to design an optimal rate in the detection of *P. velutinum* by GP are listed below:

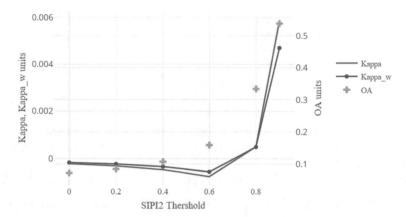

Fig. 5 The thresholding of the SIPI2 values shows how, from thresholds greater than 0.8, minimal changes are possible in the presence of different foliages in some plants, which does not imply that it is *P. velutinum* specifically. OA values reach a range of 0.5 because the fund counts as correctly classified securities

Terminal set. The implemented terminals refer to the bands captured by the multispectral camera: Red (R), Green (G), Blue (B), RedEdge (REG), Near Infrared (NIR).

Set of functions. The implemented functions include the range of arithmetic functions such as addition, subtraction, multiplication, division, and logarithms.

Population initialization. The initial population was generated in a search space under the *rampedinit* Ramped Half-and-Half method, where half of the population is initialized with the *full* method and the other with the *Grow* method, obtaining a population that is half balanced and half unbalanced, reflecting a high level of diversity. In the *full* method, each tree generates internal nodes until it reaches the assigned initial depth, and in the last depth level, it assigns terminal nodes (leaves), the individuals of this method are balanced and with branches of the same size. On the other hand, the *Grow* method randomly integrates a terminal or a function at each node, varying its depth between 1 and the maximum level, resulting in unbalanced trees.

Fitness function. The proposed function to evaluate performance is weighted Kappa (K_w), which is given by Eq. (2). This performance measure is used to evaluate the classification between evaluators, since it is a weighted estimate of the agreement between classes, evaluating for this case *n* individuals in 2 categories that are exclusive between them (presence or absence of *P. velutinum*).

Since it is a binary classification, a 2×2 confusion matrix is used in which the rows represent the real values (*Ground truth*) while the columns represent the prediction values. The stored values are true positives—*TP*, false positives—*FP*, false negatives—*FN*, and true negatives—*TN* (Banzhaf et al. 1997).

- *TP*—It is a pest and was classified as a pest
- *FP*—Not a pest but classified as a pest

- *FN*—Yes it is a pest but it was classified as not a pest
- *TN*—Not a pest and classified as a non-pest

$$K_w = 1 - \frac{\sum_{i-1}^{k} \sum_{j-1}^{k} w_{ij} x_{ij}}{\sum_{i-1}^{k} \sum_{j-1}^{k} w_{ij} m_{ij}} , \qquad (2)$$

where k is the number of categories, w_{ij} are the elements of the weight matrix, x_{ij} refers to the observed matrix and m_{ij} are the values in the matrix expected (Stehman 1996; Cohen 1968).

The weight matrix penalizes erroneous predictions, in this case, the *FP* will be given double weight than the *FN* $\begin{bmatrix} 0 & 2 \\ 1 & 0 \end{bmatrix}$, since it is more expensive in terms of time, financial and human resources to classify an area with mistletoe infestation, mobilize a commission to carry out the appropriate forest health activities and find that there is indeed no presence of the mistletoe. plague.

Selection criteria. For the selection of individuals, the type of sampling selected was the method *lexictour* (Luke and Panait 2022), which consists of the random selection of individuals that will be compared with the rest of the population, from which the best are chosen. In this comparison, only which individual is better is evaluated in parallel with another. However, if two individuals obtain the same performance value, the one chosen will be the one that has the smallest tree, that is, it has fewer nodes in-depth, implying that it will be a simpler solution to implement. On the other hand, the selected elitism method is *Keepbest* giving survival priority to the *Parents* and *children* with the greatest cliff in the weighted Kappa function, while the rest of the population follows an order in which parents come first and children second.

Data set: The proposed data set consists of 250 multispectral images (bands: R, G, B, REG, NIR) of size 300 x 300 px with the presence of *P .velutinum* and 250 segmented images (binary) of size 300 x 300 px with presence or absence of the pest, images that were preprocessed as mentioned in Sect. 3.2 (Fig. 6).

3.3.1 Experimental Design

The objective of the implementation of GP in this investigation is to find a computer program—*a spectral index* that allows the identification of *P. velutinum* on multispectral images. For this, the parameters mentioned in Table 3 are defined, where each experiment was carried out 30 times independently with different training sets. At the end of the 50 proposed generations with *kfold(5)* and implementing *Saliency Toolbox* (Walther 2006), the best individual in each experiment was evaluated with the corresponding test set to evaluate the final performance of the best candidates found. The computational time used in the execution of each iteration was 7 to 8 days for each one.

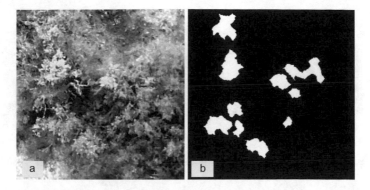

Fig. 6 *P. velutinum* Data set. **a** RGB image with the presence of *P. velutinum*. **b** Binary mask of presence/absence of the pest

Table 3 Parameters used for GP training

Generations	50
Population size	50
Initialization	Rampedinit
Crossover	Probability 0.7
Mutation	Probability 0.3
Selection	Lexictour
Elitism criteria	Keepbest
Functions set	Arithmetic operators
Terminals set	R, G, B, REG, NIR
Fitness function	Weighted Cohen's Kappa

4 Results and Discussion

From the experiments carried out, the best 30 individuals were selected (one for each experiments) and evaluated against the test data set to check their performance in the tasks of detecting the pest *P. velutinum* on multispectral aerial images. For this, the best aptitude values, the least number of nodes, and the diversity of solutions among individuals were taken into account.

Based on the diversity of functions and terminals used in GP training, it is important to discriminate their operating frequency, to study the search space, and analyze the best individuals. Figures 7 and 8 show the frequency of use in the best individuals of the terminals and functions respectively.

The test of the individuals was carried out through the confusion matrix, where additional performance metrics such as (*Overall accuracy*)—OA were considered, indicating the quality of the classification, (*Recall*)-R which reflects what proportion of the relevant elements (*P. velutinum*) that have been detected successfully, Precision—P as the predictive value positive, Cohen's kappa coefficient K and the

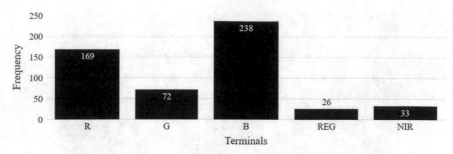

Fig. 7 Terminals frequency. The most used terminals are the Blue band and the Red band

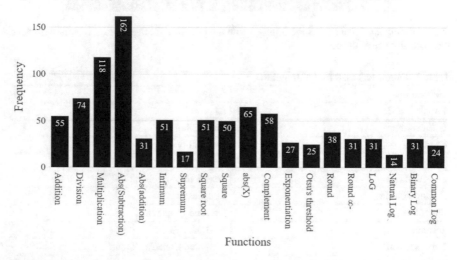

Fig. 8 Functions frequency. The most used functions are the absolute value of subtraction, multiplication, and division

Table 4 Comparison of metrics in the test and training sets—Best GP individuals

Training	Test					Diff.
K_w	K_w	k	OA	R	P	
0.4474	**0.4576**	**0.4311**	**0.9667**	**0.4086**	**0.5661**	0.0102
0.3738	0.4199	0.4082	0.9636	0.4288	0.4860	0.0461
0.3907	0.4137	0.394	0.9625	0.3972	0.5031	0.0230
0.3767	0.4084	0.3998	0.9644	0.4292	0.4737	0.0317
0.3517	0.3991	0.3939	0.9581	0.4518	0.4503	0.0474

fitness function defined for this problem *Weighted Cohen's Kappa—K_w*, the average of the metrics in the test set of the best 5 individuals is shown in the Table 4.

Given the results of the performance metrics of the best GP individuals, the best individual reports a fitness value on the test set of $K_w = 0.4576$ demonstrating greater generalizability to the detection problem of *P. velutinum* on aerial imagery while meeting an overall classification accuracy of approximately 97%.

Table 5 Comparison of metrics in the test and training sets—SIPI2 Index

Training	Test					Diff.
K_w	K_w	k	OA	R	P	
0.0016	−0.0014	0.0984	0.4953	0.5106	0.0356	−0.0030

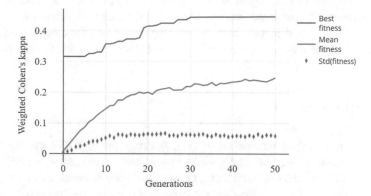

Fig. 9 Evolution of GP's individuals for the optimization of the best fitness and average fitness

On the other hand, the evaluation of the SIPI2 index was carried out on the same set of training and test data used by the best individual obtained from GP (Table 5). We can see that the fitness value obtained by this index is −0.0014 with an overall accuracy of 0.49, this is because when classifying large proportions of the image as mistletoe (because SIPI2 allows us to see changes in the health states of the tree, not necessarily caused by *P. velutinum*) increases the chances of hitting the search space; however, a large percentage of the classification is incorrect as shown by the value of k_w.

Figure 9 shows the evolution of the optimization of individuals through the generations, showing a high rate of convergence in the first 20 generations and a stabilization of the best fitness in generation 30. Average fitness throughout all generations is on average 0.3 units down throughout the evolution with a generalized standard deviation from generation 10. It is important to note that across the evolutionary process it is evident that generationally there are better solutions to the problem of detection of mistletoe in multispectral imaging.

$$((abs(\mathbf{B} - (log_{10}(\mathbf{N}))^2) \cdot log_2(abs(\mathbf{B} - \mathbf{R}))) - \mathbf{G} \cdot$$
$$(sup\left(\sqrt{\lfloor \subset (abs(\mathbf{B} - \mathbf{R}) \cdot log_2(\mathbf{N}))\rfloor}, \frac{\lfloor \subset (round(\subset (\mathbf{R}))^2)) \cdot Log_2(\mathbf{N})\rfloor}{\mathbf{B}}\right) \cdot$$
$$(abs(\mathbf{B} - \mathbf{R}) \cdot Log_2(\mathbf{G}))) ,$$

(3)

As a result of the test stage, the spatialization of the successes and errors is obtained for both the best GP individual and the SIPI2 index, as shown in Fig. 9, in the first column the original sample images, in the second the classification performed by the best individual (the best spectral index found as a solution to the detection of *P. velutinum* with GP), the third column shows the SIPI2 index throughout the thresholds and the last column shows the classification performed with SIPI2 for *P. velutinum* at a threshold of 0.8 (previously defined in Sect. 3.2).

The well-classified background can be seen in black, the *P species in white. velutinum* correctly classified, the magenta color represents the commission errors (where mistletoe was detected but there was not) and finally, the omission errors are represented in green (where there is mistletoe but it was not correctly classified).

5 Conclusions

In this research, the implementation of Genetic Programming—GP was carried out for the detection of the parasitic species *Phoradendron Velutinum*, reaching as a whole (considering the entire surface that does not refer to mistletoe) an accuracy greater than 96%. This classification was carried out employing the execution of the best evolved spectral index (individual).

From the results of training, testing, and review of the classification with the best algorithms obtained with GP, it is possible to see that in the generalization of the Phoradendron velutinum detection task on multispectral aerial images, a fitness value of 0.45 is reported but in particular cases, results of up to 0.81 are achieved in Weighted Kappa with a Recall of 0.69 and Precision of 0.89. These results and the visual analysis performed on the best and worst classifications indicate that the assigned weight matrix has a great influence on some misclassified mistletoe shoots, but also indicates that errors of omission can be caused by residuals in the registration process. between the bands of the images, additionally coupled with an error threshold found at the edges of the mistletoe since its extensions (branches and leaves) intertwine with the structure of the host tree, making it impossible to define the limit of the parasitic plant.

On the other hand, it is worth mentioning that SIPI2's approach is to evaluate the different types of chlorophyll in plants to estimate the different types of canopies and foliage, identifying certain deficiencies in the plants. However, since *P. velutinum* is a parasitic plant, it is difficult to discriminate between the vitality and the foliage of the parasite and its host; that is, the SIPI2 index cannot be used for the specific detections of *P. velutinum*.

6 Future Work

The GP technique demonstrates its versatility and adaptation for the study of different territorial problems such as the problem of forest pests addressed in this work. However, it is important to mention that given the high heterogeneity of forest areas such as the Conservation Land of Mexico City, there are still sections for improvements in the workflow and parameters considered here. It is important to delve into the use of the proposed methodology because despite reaching an average Overall Accuracy of around 96% in the first experiments, several misclassification errors were overlooked in which mistletoe was not correctly detected, generally in areas close to those already correctly classified due to the irregular shape of the parasitic species. On the other hand and more importantly, there are several errors in which it was classified as *P. velutinum* but in reality, this pest did not exist in that area, which is reflected in an average Precision of 56%. It is important to mention that these *commission* errors can present a high cost in forest sanitation activities.

Additionally, it should be noted that it is necessary to deepen the study and analysis of better individuals, for this, it is necessary to continue with a greater number of experiments, which implies a greater investment in processing time.

From the development of this work, several questions of improvements arise both at the level of the acquisition of images, recording of additional spectral bands, the study of the best functions, the best parameters, and the evaluation of performance compared to other algorithms of machine learning classification.

Acknowledgements The authors would like to thank Ing. Israel Jimenez and the Commission for Natural Resources and Rural Development—CORENA for the support and contribution of their experience and accompaniment in the field for the detection of *P. velutinum* in the conservation soil of the City of Mexico.

References

Alvarado-Rosales D, Equihua-Martínez A, López-Gomez Tagle E, Rodríguez-Ortega A, de Lourdes Saavedra-Romero L, Vanegas-Rico JM (2007) Situación actual de la declinación del oyamel en el parque Desierto de los Leones, D.F. In: Memoria del XIV Simposio Nacional de Parasitología Forestal. Instituto Nacional de Investigaciones Forestales, Agrícolas y Pecuarias, pp 24–28

Banzhaf W, Keller RE, Francone FD, Morgan MB (1997) Genetic programming: an introduction. Morgan Kaufmann

Bay H, Tuytelaars T, Van Gool L (2006) SURF: Speeded up robust features. In: Leonardis A, Bischof H, Pinz A (eds) Computer vision - ECCV 2006. Springer, Berlin, Heidelberg, pp 404–417

Bi Y, Bing X, Mengjie Z (2021)Genetic programming for image classification. Springer Nature

Brameier M, Banzhaf W (2001) A comparison of linear genetic programming and neural networks in medical data mining. IEEE Trans Evolut Comput 5(1):17–26

Chaudhary UK, Iqbal M (2009) Determination of optimum genetic parameters for symbolic nonlinear regression-like problems in genetic programming. In: INMIC 2009 - 2009 IEEE 13th international multitopic conference, vol 042, pp 12–16

Chen J, Li ZZ, Liao ZG, Wang YL (2005) Distributed service performance management based on linear regression and genetic programming. In: 2005 international conference on machine learning and cybernetics. ICMLC 2005, pp 560–563

Cohen J (1968) Weighted kappa: nominal scale agreement with provision for scaled disagreement or partial credit. Psychol Bullet 70(4):213–220

Cramer NL (1985) A representation for the adaptive generation of simple sequential programs. In: International conference on genetic algorithms and the applications, pp 183–187

Crow FC (1984) Summed-area tables for texture mapping. Comput Gr (ACM) 18(3):207–212

Darwin C (1859) On the origin of species by natural selection

Dwivedi R, Dey S, Chakraborty C, Tiwari S (2021) Grape disease detection network based on multi-task learning and attention features. IEEE Sens J 21(16):17573–17580

Goncalves I, Silva S (2013) Balancing learning and overfitting in genetic programming with interleaved sampling of training data. In: Krawiec K, Moraglio A, Hu T, Etaner-Uyar AŞ Hu B (eds) Genetic programming. EuroGP, vol 7831. LNCS (December 2016)

Gonzáles Gaona E, Serrano Gómez C, De Lira Ramos KV, Quiñonez Barraza S, Sánchez Martínez G, López Pérez I, Sánchez Lucio R (2017) Identificación, distribución y control de muérdago enano (Arceuthobium spp.) en bosques de coníferas. Instituto Nacional de Investigaciones Forestales, Agrícolas y Pecuarias, Aguascalientes

Gutiérrez Vilchis LH, Reséndiz-Martínez JF (1994) Fenología del muérdago enano en el desierto de los leones, D.F. Revista de Ciencia Forestal en México 19(75):41–62

Huan-rong Z, Ya-min L, Ia-mei M (2010) Based on meteorological factors and short-term load forecasting genetic programming. In: 2010 International conference on computer design and applications, vol 3. IEEE, pp 465–467

Huo L, Fan X, Xie Y, Yin J (2007) Short-term load forecasting based on the method of genetic programming. In: Proceedings of the 2007 IEEE international conference on mechatronics and automation, ICMA 2007, pp 839–843

Icke I, Bongard JC (2013) Improving genetic programming based symbolic regression using deterministic machine learning. In: 2013 IEEE congress on evolutionary computation, CEC 2013, pp 1763–1770

Koza JR (1992) Genetic programming: on the programming of computers by means of natual selection. Massachusetts

Koza JR (1994) Genetic programming as a means for programming computers by natural selection. Stat Comput 4(2):87–112

Lee HS, Seo WW, Lee KS (2019) Detection of Oak Wilt disease using convolutional neural network from Uav natural color imagery. In: IGARSS 2019 - 2019 IEEE international geoscience and remote sensing symposium. IEEE, pp 6622–6624

León Bañuelos LA (2019) Análisis de la distribución espacial de Arceuthobium Globosum implementando teledetección en el área de protección de flora y fauna Nevado de Toluca. PhD thesis, Universidad Autónoma del Estado de México

León-Bañuelos LA, Endara-Agramont AR, Gómez-Demetrio W, Martínez-García CG, Nava-Bernal EG (2020) Identification of Arceuthobium globosum using unmanned aerial vehicle images in a high mountain forest of central Mexico. J For Res 31(5):1759–1771

Lowe DG (1999) Object recognition from local scale-invariant features. In: Proceedings of the international conference on computer vision, vol 2, pp 1150–1157

Luke S, Panait L (2020) Lexicographic parsimony pressure. In: GECCO 2002: proceedings of the genetic and evolutionary computation conference, pp 829–836

Mathiasen RL, Nickrent DL, Shaw DC, Watson DM (2008) Mistletoes: pathology, systematics, ecology, and management. Plant Disease 92(7):988–1006

MATLAB (2019) Image processing toolbox ^{TM}User ' s guide R 2019 b. Technical report, The MathWorks, Inc

Mihaylov R, Atanasow A, Ivanova A, Marinov A, Zahariev S (2020) Tracking the Development of Six Wheat Varieties Using Infrared Imaging and Image Processing Algorithms. *2020 International Conference Automatics and Informatics, ICAI 2020 - Proceedings*, 2020

Minařík R, Langhammer J, Lendzioch T (2020) Automatic tree crown extraction from UAS multispectral imagery for the detection of bark beetle disturbance in mixed forests. Remote Sens 12(24):1–31

Morales AK, Casas JG (2007) Algoritmos genéticos. Sociedad Mexicana de Inteligencia Artificial, Ciudad de México

Peñuelas J, Filella I (1995) Reflectance assessment of mite effects on apple trees. Int J Remote Sens 16(14):2727–2733

Peres DJ, Cancelliere A (2021) Analysis of multi-spectral images acquired by UAVs to monitor water stress of citrus orchards in sicily, Italy. In: World environmental and water resources congress, pp 270–278

Pernar R, Bajić M, Ančić M, Seletković A, Idžojtić M (2007) Detection of mistletoe in digital colour infrared images of infested fir trees. Period Biol 109:67–75

Poli R, Langdon WB, McPhee NF, Koza JR (2008) A field guide to genetic programing

Qi H, Wu Z, Zhang L, Li J, Zhou J, Jun Z, Zhu B (2021) Monitoring of peanut leaves chlorophyll content based on drone-based multispectral image feature extraction. Comput Electron Agric 187(June):106292

Rzedowski J, de Rzedowski GC (2011) Flora Del Bajío Y De Regiones Adyacentes. Instituto de Ecología 170:222–235

Sabrina F, Sohail S, Thakur S, Azad S, Wasimi S (2020) Use of deep learning approach on UAV imagery to detect mistletoe infestation. In: 2020 IEEE region 10 symposium, TENSYMP 2020, pp 556–559

Schmitt R (2002) 1 - introduction and survey of the electromagnetic. In: Electromagnetics explained. Newnes, pp 1–24

Shankar RLH, Veeraraghavan AK, Sivaraman K, Ramachandran SS (2018) Application of UAV for pest, weeds and disease detection using open computer vision. In: Proceedings of the international conference on smart systems and inventive technology, ICSSIT 2018, (Icssit), pp 287–292

Silva D, Almeida J (2007) GPLAB a genetic programming toolbox for MATLAB. In: Proceedings of the nordic MATLAB conference

Stehman SV (1996) Estimating the kappa coefficient and its variance under stratified random sampling. Photogram Eng Remote Sens 62(4):401–407

Thigpen J, Shah SK (2008) Multispectral imaging. In: Microscope image processing, 4th edn., pp 299–327

Universidad Nacional Autónoma de México (2022) Departamento de Botánica, Instituto de Biología (IBUNAM), Phoradendron velutinum (DC.) Oliv., ejemplar de: Herbario Nacional de México (MEXU), Plantas Vasculares

Walther D (2006) Interactions of visual attention and object recognition: computational modeling, algorithms, and psychophysics. PhD thesis, California Institute of Technology

Xie Q, Huang W, Liang D, Chen P, Wu C, Yang G, Zhang J, Huang L, Zhang D (2014) Leaf area index estimation using vegetation indices derived from airborne hyperspectral images in winter wheat. IEEE J Sel Top Appl Earth Obs Remote Sens 7(8):3586–3594

Xue SY, Xu HY, Mu CC, Wu TH, Li WP, Zhang WX, Streletskaya I, Grebenets V, Sokratov S, Kizyakov A, Wu XD (2021) Changes in different land cover areas and NDVI values in northern latitudes from 1982 to 2015. Adv Climate Change Res

Yeom J, Jung J, Chang A, Maeda M, Landivar J (2017) Cotton growth modeling using unmanned aerial vehicle vegetation indices. In: IEEE international geoscience and remote sensing symposium (IGARSS), pp 5050–5052

Yuan Y, Fan W, Wang W, Liu H (2008) Robust collaborative optimization of a Multi-finger Micro-accelerometer based on genetic algorithm. In: Proceedings - 2nd international conference on genetic and evolutionary computing, WGEC 2008, vol 60474059, pp 105–108

Zhang N, Wang Y, Zhang X (2020) Extraction of tree crowns damaged by Dendrolimus tabulaeformis Tsai et Liu via spectral-spatial classification using UAV-based hyperspectral images. Plant Methods 16(1):1–20

Assessment of the Reduction of the Icesnow Coverage at the TransMexican Volcanic Belt Through Empirical Mode Decomposition on Satellite Imagery

Alfredo Sánchez-Martínez, Emiliano Yahel Ruíz-Oropeza,
Mauricio Gabriel Orozco-del-Castillo, Jorge J. Hernández-Gómez,
and Gabriela Aurora Yáñez-Casas

Abstract The drastic decrease in snow and ice cover on the main peaks of the Trans-Mexican Volcanic Belt has been observed by both the civilian population and scientists. This decrease has occurred particularly in the once considered permanent glaciers on such peaks. Here, Landsat images are used to evaluate the change in snow and ice cover on the three main peaks of the Trans-Mexican Volcanic Belt: Iztaccíhuatl (5,230 m.a.s.l.), Popocatépelt (5,426 m.a.s.l.) and Citlaltépetl (5,636 m.a.s.l). Through image segmentation techniques, we obtain a time series of snow and ice cover for each volcano. Subsequently, temporal tendencies were obtained through the empirical mode decomposition technique. For the study period between 1985 and 2020, the analysis show a clear decreasing trend in the area covered by snow and ice (an average reduction of 27.74% for Iztaccíhuatl and Citlaltépetl). This is particularly intense for Popocatépetl volcano (99.93 % from 1985 to 2020). The behaviour of the snow and ice cover time series seems to be attributed to some periodicities in the solar cycle, as well as the eruptive activity of Popocatépetl volcano since 1999; another potential cause could be the impact of global warming in the region.

A. Sánchez-Martínez
Instituto Politécnico Nacional, Escuela Superior de Cómputo, Av. Juan de Dios Bátiz S/N, Lindavista, Gustavo A. Madero, 07738 Ciudad de México, Mexico

A. Sánchez-Martínez · E. Y. Ruíz-Oropeza
Instituto Politécnico Nacional, Centro de Estudios Científicos y Tecnológicos No. 9 Juan de Dios Bátiz, Mar Mediterráneo 227, Popotla, 11400 Ciudad de México, Mexico

M. G. Orozco-del-Castillo
Tecnológico Nacional de México / IT de Mérida, Departamento de Sistemas y Computación, Av. Tecnológico km. 4.5 S/N, 97118 Mérida, Yucatán, Mexico

J. J. Hernández-Gómez (✉) · G. A. Yáñez-Casas
Instituto Politécnico Nacional, Centro de Desarrollo Aeroespacial, Belisario Domínguez 22, Centro, 06010 Ciudad de México, Mexico
e-mail: jjhernandezgo@ipn.mx

1   Introduction

Diverse phenomena lead to variations in the extension of the snow and ice fields that cover the high peaks of the mountains. Such phenomena have different origins (e.g., natural or anthropogenic), producing important and drastic changes in the Earth systems. In recent decades, the effects caused by climate change have had the greatest impact in the reduction of both the superficial ice and the permafrost (Cong et al. 2020; Liu et al. 2017a; Rangecroft et al. 2016; Yokohata et al. 2021).

In volcanic regions, the associated geological processes influence the extent of the snow and ice alongside the regional climate. Geological processes include the exogenous processes of erosion and accumulation of rock debris due to volcanism (Keller 2008). The effects of pluvial and snow precipitations also favour erosion processes and affect the accumulation of snow and ice (Lugo-Hubp et al. 1981); however, these natural processes are not the main cause of the rapid decrease in snow and ice cover on the tops of volcanoes.

Currently, throughout the world, large ice structures are retreating, mainly due to the effects of climate change, which causes atypical reactions and changes in natural cycles (e.g. water and temperature cycles) (Nguvava et al. 2019; Yadollahie 2019). Specifically, the increase in the temperatures of the previously frozen areas leads to drastic melting of the snow and ice cover (Mojica Moncada et al. 2020), the decrease in the glacial snow layers (Moreno et al. 2020), as well as the deformations in the geological structure of mountain systems (Ames and Hastenrath 1996).

The peaks of the Trans-Mexican Volcanic Belt (TMVB) are no exception (White 1981). There are few surfaces that maintain positive balances between the processes of melting and recovery of ice and snow. This condition occurs in places where the climate favours the conservation of the balance. In addition to surface area reduction, the snow and ice fields also show thinning, which implies a loss of mass (Barth et al. 2019; Song et al. 2017). The reduction of the snow and ice coverage has accelerated since 1968 due to global climate changes (White 2002). Likewise, atypical behaviours have been observed in periglacial activity (freezing and thawing) at altitudes above 4,600 m.a.s.l. (Heine 1975, 1994; Villalpando and Ik 1968).

Of the volcanoes in the TMVB, it is known that the Iztaccíhuatl had 12 glaciers up to 1959 (Lorenzo 1959). By 1982, some of these glaciers ("del cuello", "oeste-noroeste", "suroriental" and "San Agustín") had already disappeared (White 2002), reducing the area covered by glaciers from 1.2 km^2 in 1959 to 1 km^2 in 1982. On the other hand, by Delgado-Granados (1996), Popocatépetl had an area covered by glaciers of approximately 0.5 km^2, of which the "ventorrillo" and "noroeste" glaciers are the ones that have been most affected by environmental pollution from Mexico City and the increase in temperature that this entails (Delgado-Granados 1996). Finally, with respect to the Citlaltépetl volcano (or Pico de Orizaba), where vestiges of the advances of the Last Glacier Maximum of the Little Ice Age (~1250–1830 CE Pompeani et al. 2021) are preserved, a significant retreat in ice covered area was observed. Since the 1950s, the lower limit of the glacier has significantly moved, for instance, from 4,640 m.a.s.l. at the beginning of 20th century to 4,980 m.a.s.l. in 1958

(Cortés-Ramos and Delgado-Granados 2015). The receding dynamics left blocks of ice abandoned in the lower part of the valley of "la canaleta" and "los laberintos". In addition, hillside processes (cryoclasts and smaller diameter sediments as moraine deposits composed of pumice and ash) have buried some of these ice remnants (Soto-Molina et al. 2019).

Since the decrease of glaciers can trigger a series of adverse events, it is important to focus efforts to mitigate it. Computational techniques, specially those based on artificial intelligence, play an essential role in the interpretation of field data because of their easiness of implementation, flexibility, and reduced computational cost (Dikshit et al. 2020; Russell et al. 2015). In this sense, techniques such as fuzzy theory (Gajek et al. 2017), agent-based modelling (Terzi et al. 2019), swarm optimisation (Rahman et al. 2013), genetic algorithms (Velez et al. 2011), pattern recognition (Quincey et al. 2007), neural networks (Wang et al. 2021) and machine learning (Al-Najjar et al. 2019) have been used in the study of the reduction of glaciers both on volcanic peaks and other areas.

Using image segmentation techniques, in this study we show a time series analysis of the accumulation and retreat of the ice and snow cover on the peaks of the highest volcanoes in the TMVB: Iztaccíhuatl, Popocatépetl, and Ciltlaltépetl. Using Empirical Mode Decomposition (EMD), we observe a clear decreasing trend in the accumulation of snow and ice and a significant decrease of the surface area covered by snow and ice from 1985 to 2020. The present work is organised as follows: in Sect. 2 we present the methods adopted and the study area. In Sect. 3 we present the results. In Sects. 4 and 5 we present the discussion of the results and the conclusions, respectively. Possible future work is also presented.

2 Methods and Field Study Area

2.1 Methods

2.1.1 Thresholding Techniques

Segmentation techniques are fundamental for the detection of objects in digital images (Mardia and Hainsworth 1988). In this work, we consider the problem of digital image segmentation by thresholding technique (Gonzalez and Woods 2008). If the value of an image f at (x, y) is given by $f(x, y)$, where x and y are the coordinates of the pixels, in which light objects are placed on a dark background, or vice versa, then a way to extract sub-objects is to set a threshold T that separates both objects and background. Then, any pixel (x, y) at which $f(x, y) > T$ is recognised as an object point; otherwise, it is recognised as a background point. According to Gonzalez and Woods (2008) the intensity thresholded image $(g(x, y))$ is given by the equation:

$$g(x, y) = \begin{cases} 1, & \text{if } f(x, y) > T \\ 0, & \text{if } f(x, y) \leq T \, . \end{cases}$$

Global thresholding occurs when T is constant over the whole image. Moreover, multiple thresholding occurs if several thresholds are set in order to distinguish objects with more than three intensities within the image. Thresholding has been a widely used technique in science and technology (Hernandez-Capistran and Martinez-Carballido 2016; Houssein et al. 2021; Pare et al. 2020).

2.1.2 Kriging Method

The Kriging method is an advanced interpolation procedure, which aims to build an approximate surface from a set of points that are separated from each other. This procedure is based on the adjustment of a mathematical function taking into account the specified points, or delimiting those points within a specific radius to determine the output value for each location. This algorithm assumes that the distance or direction between the sample points indicates a spatial correlation that can be used to generate a surface with suitable variations. Thus, the Kriging method is useful when there is a directional influence on the spatially correlated distance within the data (Orozco-del-Castillo et al. 2019).

For the Kriging method to consider and evaluate the surrounding points measured to calculate a certain prediction of a location without specific measurements, a geostatistical estimator is used (Suarez-Gallareta et al. 2018). The Kriging method has been broadly used in natural sciences, applied mathematics and computer science (Liu et al. 2017b; Zhang et al. 2017).

2.1.3 Empirical Mode Decomposition

EMD was first presented by Huang et al. (1998). This algorithm produces soft envelopes defined by local ends of a sequence and the subsequent subtraction of the mean of these envelopes from an initial sequence. It requires identifying all the local endpoints connected by cubic spline lines to produce the upper and lower envelopes (Kim and Oh 2009). EMD has been extensively used in different science fields to perform operations as recognition (Du et al. 2015), analysis (Nunes et al. 2003), filtering (Andrade et al. 2006) and prediction (Drakakis 2008).

This adaptive analysis method is suitable for processing non-stable and non-linear time series. The ultimate purpose of EMD is to divide the series into modes $D_i(t)$, the so-called Intrinsic Mode Functions (IMFs) in the time domain. The latter constitutes the main difference between EMD and other analysis methods for time series such as the Fourier transform and the decomposition into wavelets, which perform a time domain transformation. Since EMD always remains in the time domain, it has the advantage of taking less processing time to produce results. Further information about the EMD method can be found in the work by Huang (2001).

2.2 Case Study Region

2.2.1 Geographical Setting

The case study region is located within the TMVB (Fig. 1), a volcanic arc that covers Cretaceous and Cenozoic magmatic provinces (Ferrari et al. 2012). It is a region related to extensional and normal faults in the subducted Cocos and Rivera plates under the North-American plate (Johnson and Harrison 1990; Pasquarè et al. 1987). The TMVB is located within the coordinates 18°30' and 21°30', and extends from the coast of the Pacific Ocean to the Gulf of Mexico (Pérez-Moreno et al. 2021). It has an approximate length of 1,000 km and a variable width of 90–230 km (Pérez-Moreno et al. 2021).

Here, we look at the three highest volcanic peaks of the TMVB: Iztaccíhuatl (5,230 m.a.s.l.), Popocatépelt (5,426 m.a.s.l.) and Citlaltépetl (5,636 m.a.s.l). These

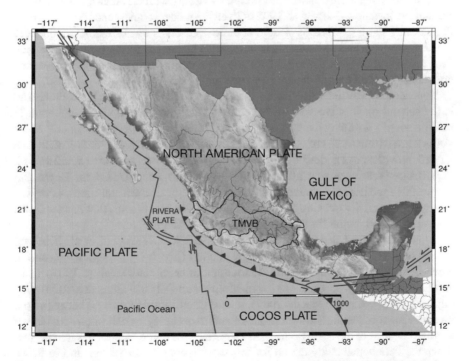

Fig. 1 Location of TMVB, along with the interaction between Cocos, Rivera and North American tectonic plates. Image redrawn from Pérez-Moreno et al. (2021), which is published under a CC BY 4.0 licence

Table 1 Coordinates of the region under study

Volcano	Top-left	Bottom-right
Iztaccíhuatl	N 19° 14'	N 19° 06'
	W 98° 34'	W 98° 42'
Popocatépetl	N 19° 06'	N 18° 58'
	W 98° 42°	W 98° 34'
Citlaltépetl	N 19° 06'	N 18° 58'
	W 97° 20'	W 97° 11'

volcanoes are located between 18 ° 30'–20 ° 15' and 98 ° 20'N–100 ° 20'W. The origin of the main landforms is the volcanism of the Neogene-quaternary, as well as the tectonism that created folded mountains in the middle-late Tertiary. Both give rise to diverse land surfaces that favour exogenous erosion and accumulation processes of rock debris on slopes (Lugo-Hubp et al. 1981).

Iztaccíhuatl volcano (5,230 m.a.s.l.) is a dormant stratovolcano (Macías et al. 2012); its last volcanic manifestation was the formation of the Teyotl dacitic dome, about 80 ka ago (Nixon 1989). However, flank activity at the southern edge of the complex occurred during the Holocene about 9 ka ago, with the emission of the dacite Buenavista lava flow (Siebe et al. 1995). Given the volcanic quiescence during the late Pleistocene, the Iztaccíhuatl volcanic complex preserves one of the best sequences of moraine deposits in Mexico. It is is located, 60 km SE of the centre of Mexico City and 50 km NW of the city of Puebla. The coordinates that narrow down the Iztaccíhuatl region for this study are shown in Table 1.

Popocatépetl volcano (5,426 m.a.s.l.) is located only 16 km from Iztaccíhuatl and is the second most active volcano in Mexico, after the Colima volcano. About 27 million people, 8 million dwellings, 4,847 health centres, 45 thousand schools and 7 airports are within a 120 km radius from Popocatépetl (Rodríguez-Pérez et al. 2021). It is a stratovolcano of andesitic composition with a summit elevation of 5,452 m.a.s.l. Popocatépetl reawakened in 1994 after 70 years of dormancy (De la Cruz-Reyna and Siebe 1997) and has been characterised by recurring gas and ash emissions, as well as by explosive activity (Mendo-Pérez et al. 2021). The coordinates that narrow down the Popocatépetl region for this work are shown in Table 1.

Citlaltépetl (5,636 m.a.s.l) is a dormant stratovolcano (Rossotti et al. 2006) on the Eastern TMVB, located 200 km from Mexico City. It is the highest volcano in North America and the third highest mountain after Mount Denali (6,190 m.a.s.l.) and Mount Logan (5,959 m.a.s.l.) (Soto-Molina and Delgado-Granados 2020). Citlaltépetl age is ~650 ka (Macías 2007). Its last eruptive phase is dated between 16.5 ka and 4 ka ago, yielding to recent dacite and andesite pyroclastic flows that cover large extensions of the north of the cone (Macías 2007). The coordinates that narrow down the Citlaltépetl study region for this work are shown in Table 1. In Fig. 2, we show the areas of study for each of the three volcanoes.

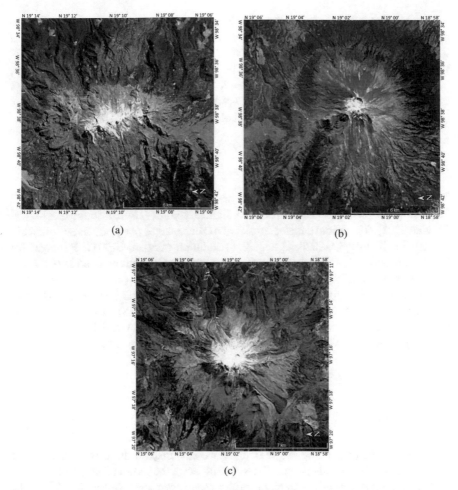

Fig. 2 Satellite images showing the areas of interest for the three volcanoes: **a** Iztaccíhuatl volcano, **b** Popocatépetl volcano, and **c** Citlaltépetl volcano. These images were obtained from United States Geological Service (2021) and were taken on July 30th, 1986

The region comprising the studied volcanoes is characterised by having purely tropial climatic conditions, with little annual thermal oscillation. The rainfall is concentrated in summer, during the months of May to October due to the dominance of the inter-tropical convergence zone (Andrés et al. 2010). The winter season is identified with the predominance of atmospheric stability corresponding to the high-pressure subtropical zone. Occasionally, scarce precipitations during this season are caused by polar air masses (mainly continental) (Andrés et al. 2010). The dry season occurs between December and February. The meteorological stations with altitudes around 2,300 m.a.s.l. indicate an Annual Mean Air Temperature (AMAT) of 16 °C and rainfall of 850 mm per year. Beaman (1962) has computed

8 °C AMAT at 3,500 m.a.s.l. and 5 °C AMAT at 4,000 m.a.s.l. (Beaman 1962) also assumes that precipitations initially rises with altitude, with 1,000 mm per year at 2,700 m, 1,200 mm at 3,000 m and 1,300 mm at 3,200 m. From here, the precipitation decreases to 1,200 mm at 4,000 m. The altitude limit of the forest is, on average, 3,900 m (Beaman 1962). Above this limit the alpine pasture floor extends, with the formation of tall grasses, and from 4,400 m the vascular vegetation disappears (Almeida et al. 1994; Almeida-Lenero et al. 2004).

2.2.2 Satellite Imagery

The Landsat satellite images used in the present study span in time from 1985 until 2020 and were downloaded from USGS Earth Explorer (United States Geological Service 2021). The initial dataset consisted of four images per year, taken from July 30, 1985 until July 30, 2020, at a spatial resolution of 30 m. GEOTIFF images for layered analysis were obtained from Landsat 8 and 7 Collections I and II (CL1 and CL2, respectively), levels 1 and 2 (L1 and L2, respectively).

Within this time span, the following sensors acquired the images: the Landsat Operational Land Imager/Thermal Infrared Sensor (OLI/TIRS) CL2 L1 and L2, the Landsat Enhanced Thematic Mapper Plus (ETM+) CL2 L1 and L2, the Landsat Thematic Mapper (TM) C2 L1 and L2, as well as the Multispectral Scanner System MSS C2 L1.

3 Results

In order to obtain a reliable set of images, we first discard those with large cloud coverage which would otherwise set false positives for the snow and ice identification. The set of images comprises elements with a cloud coverage less than 5%. Then, the images were manually selected in order to guarantee that the cloud coverage was not intense enough to hinder the identification of snow and ice. In Table 2, the dates of the final dataset are shown, for each of the three volcanoes.

For the purpose of determining the optimal threshold value to segment the image between both snow and ice, and background classes, values of the threshold T between 150 and 255 were tested. Two experts were consulted in order to determine that the value of threshold that yielded best segmentation results for this problem is $T = 240$. The results of the segmentation with this threshold value for the satellite images of 1986 (Fig. 2) are shown in Fig. 3.

To have a normalised measurement, the number of pixels covered by snow and ice was divided by the total number of pixels forming the image. Next, using the Kriging method, we interpolate the dates of the images in order to obtain the same length and time series for all three volcanoes. In total, each interpolated time series has 42 points. The time series of snow and ice coverage for each volcano is shown in Fig. 4.

Table 2 Dates of each satellite image within the final (cloudless) dataset

Iztaccíhuatl	Popocatépetl	Citlaltépetl
12-30-1985	12-30-1985	7-30-1985
7-30-1986	7-30-1986	7-30-1986
7-30-1987	7-30-1987	7-30-1987
7-30-1988	7-30-1988	7-30-1988
7-30-1989	7-30-1989	7-30-1989
7-30-1990	7-30-1990	7-30-1990
7-30-1991	12-30-1993	12-30-1991
7-30-1992	7-30-1994	7-30-1992
7-30-1993	7-30-1995	12-30-1993
7-30-1994	7-30-1996	7-30-1994
7-30-1995	7-30-1997	7-30-1995
7-30-1996	7-30-1998	7-30-1996
7-30-1999	7-30-1999	7-30-1997
7-30-2000	7-30-2000	7-30-1998
7-30-2001	7-30-2001	7-30-1999
12-30-2003	7-30-2002	7-30-2000
7-30-2004	7-30-2003	7-30-2001
7-30-2005	7-30-2004	7-30-2002
7-30-2006	7-30-2005	10-30-2003
7-30-2007	7-30-2006	7-30-2004
7-30-2008	7-30-2007	7-30-2005
7-30-2009	7-30-2008	7-30-2006
7-30-2010	7-30-2009	7-30-2007
7-30-2011	7-30-2010	7-30-2008
7-30-2012	7-30-2011	7-30-2009
7-30-2013	7-30-2012	7-30-2010
7-30-2014	7-30-2013	7-30-2011
7-30-2015	7-30-2014	7-30-2012
12-30-2017	7-30-2015	7-30-2013
7-30-2018	7-30-2016	7-30-2014
7-30-2019	7-30-2017	7-30-2015
7-30-2020	7-30-2018	7-30-2016
	7-30-2019	7-30-2017
	7-30-2020	7-30-2018
		7-30-2019
		7-30-2020

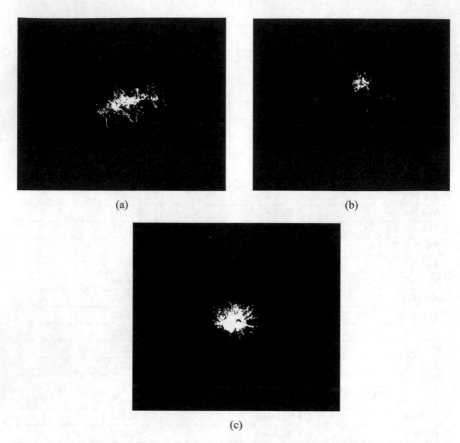

Fig. 3 Results of the post-processing analysis using the segmentation methods with $T = 240$, for the images taken on July 30, 1986, of the **a** Iztaccíhuatl volcano, **b** Popocatépetl volcano, and **c** Citlaltépetl volcano

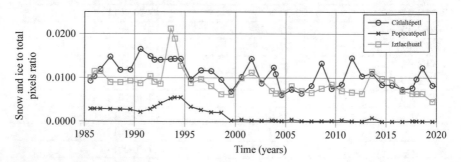

Fig. 4 Time series of snow and ice coverage obtained from the analysis of satellite images, for each of the volcanoes. The blue line represents Citlaltépetl, the red one Popocatépetl, and the green one Iztaccíhuatl

The time series of snow and ice coverage for each volcano were analysed using EMD (Fig. 4). The first four IMFs are shown in Fig. 5.

4 Discussion

In order to assess the reduction in the snow and ice coverage on the studied volcanoes, the behaviour of the first four IMFs obtained by the application of EMD to the original snow and ice coverage time series (Fig. 4), are shown in Fig. 5. Furthermore, in Table 3 we report the identified periods of such computed IMFs. These periodicities were obtained as the mean of the periods between consecutive peaks or valleys of the IMFs.

The most notable result that ought to be remarked is that the $D_1(t)$ IMF (Fig. 5a), reveals a clear decreasing trend in the snow and ice coverage from 1985 to 2020, for the three volcanoes. It is noticeable that the decreasing trend in the snow and ice coverage within the considered period is the same even for the Popocatépetl, which reawakened in 1994 and has had continuous volcanic activity. This fact can be observed in the original time series (Fig. 4), in which it is seen that the behaviour of the Popocatépetl time series departs from the behaviour of the time series for the other two volcanoes. The analysis shows a minimal snow and ice coverage starting from 1999 until present days. It is worth mentioning that Popocatépetl lost a significant percentage (99.55 %, according to our results) of its glaciers since the 2000s, in agreement with (Veettil and Wang 2018). Although the glacier decreasing trend for the three volcanoes can be attributed to pollution and climate change (Veettil and Wang 2018), the Popocatépetl glacier retreat is strongly driven by its volcanic eruptions (Mendo-Pérez et al. 2021).

As for the periodicities identified in the EMD IMFs in Fig. 5, and which are shown in Table 3, we can make the following observations.

The second IMF, $D_2(t)$ (Fig. 5b) reveals a coarse frequency that corresponds to a mean period of 19.66 years, which is concordant with bidecadal oscillation of the global surface temperature, as first noted by Ghil and Vautard (1991). This oscillation is likely due to the occurrence of the Hale cycle. Hale is a cycle with around 20 years that frequently appears in meteorological records, and in some instances, it is closely related to the double-Sunspot cycle (Burroughs 2005) Furthermore, it is interesting to note that the second maximum for such time series occurred between 2012 and 2014 (Fig. 5b), which is close to the date of the maximum of solar cycle number 23 (April, 2014) (Orozco-del-Castillo 2017), which reinforces the hypothesis that further variations revealed by the subsequent IMFs are partially driven by solar activity.

In order to identify if the periodicities found in the subsequent IMFs ($D_i(t)$ for $i = 3, 4$) are driven by solar activity, they are compared with those reported by Kane (2005) and Orozco-del-Castillo (2017). The latter records, which are obtained from different Sun activity indicators (e.g. Ground Level Enhancements (GLEs), number of Sunspots), are a set of short term periods within the solar activity (Table 4).

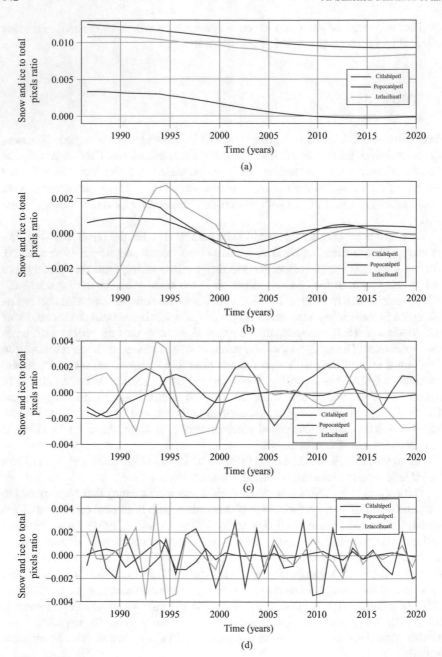

Fig. 5 Individual view of the first four Intrinsic Mode Functions (IMFs) obtained through the application of EMD to the snow and ice coverage time series shown in Fig. 4. The blue line stands for Citlaltépetl, the red line for Popocatépetl and the green line for Iztaccíhuatl. **a** $D_1(t)$. **b** $D_2(t)$. **c** $D_3(t)$. **d** $D_4(t)$

Table 3 Identified periodicities in the four computed IMFs through EMD (Fig. 5)

IMF	Mean period (year)		
	Iztaccíhuatl	Popocatépetl	Citlaltépetl
$D_2(t)$	18.00	21.00	20.00
$D_3(t)$	5.20	7.10	10.30
$D_4(t)$	3.20	3.85	3.00

Table 4 Periods of the thirty harmonics obtained from Kane (2005) and Orozco-del-Castillo (2017). The periods are given in years

Harmonic	Period	Harmonic	Period	Harmonic	Period
1	10.83	11	0.98	21	0.52
2	5.42	12	0.90	22	0.49
3	3.61	13	0.83	23	0.47
4	2.71	14	0.77	24	0.45
5	2.17	15	0.72	25	0.43
6	1.81	16	0.68	26	0.42
7	1.55	17	0.64	27	0.40
8	1.35	18	0.60	28	0.39
9	1.20	19	0.57	29	0.37
10	1.08	20	0.54	30	0.36

It is important to note that IMFs $D_3(t)$ and $D_4(t)$, have periods within the range of those reported by Kane (2005) and Orozco-del-Castillo (2017). $D_3(t)$ reveals the period of the solar cycle (Harmonic 1 in Table 4), which is entirely observable in Citlaltépetl, which is far from the influence of the volcanic activity of Popocatépetl. However, Iztaccíhuatl resembles for D_3 the periodicity contained in the second harmonic of solar activity (Table 4). In this sense, the influence of the volcanic activity of Popocatépetl is observed in $D_3(t)$, which almost averages the values obtained for Iztaccíhuatl and Popocatépetl volcanoes.

$D_4(t)$ IMF coincides for the three volcanoes, yielding an average value of 3.35 years, which is closely related to the harmonic 3 of the solar activity (Table 4). Therefore, IMF 4 appears to be fully driven by the Sun activity.

5 Conclusions

The main periodicities in the time series of snow and ice coverage records from July 30, 1985 to July 30, 2020, are presented for the three highest volcanoes in the TMVB: Iztaccíhuatl, Popocatépetl and Citlaltépetl. The analysis was conducted by applying the global thresholding segmentation technique to satellite images obtained from

Table 5 Comparison of the obtained periods through the application of EMD with those reported in the literature for the solar activity (Kane 2005; Orozco-del-Castillo 2017) for the coincident IMFs, along with the relative error between both

Volcano	IMF	EMD period (year)	Measured period (year)	Relative error (%)
Iztaccíhuatl	$D_2(t)$	18.00	20.00	10.00
	$D_3(t)$	5.2	5.42	4.05
	$D_4(t)$	3.20	3.61	11.35
Popocatépetl	$D_2(t)$	21.00	20.00	5.00
	$D_3(t)$	7.10	5.42	30.99
	$D_4(t)$	3.85	3.61	6.64
Citlaltépetl	$D_2(t)$	20.00	20.00	0.00
	$D_3(t)$	10.30	10.83	4.89
	$D_4(t)$	3.00	3.61	16.89

Landsat imagery, with minimal clouud coverage. Periodicities in the snow and ice coverage as a function of time were obtained through EMD, after the time series were interpolated through the Kriging method. The first IMF revealed a clear decreasing trend of the snow and ice coverage for the three volcanoes. This temporal tendency might be associated to effects of global warming in the region, as has been proven for glaciers in the Himalayas (Song et al. 2017).

On the other side, larger period components on IMFs $D_2(t)$ through $D_4(t)$ are most likely driven by the solar activity. In order to better grasp this fact, in Table 5 we present a summary of the EMD obtained periods against those reported in the literature. We also present the relative error between both values. Some IMFs match to the reported periods in the solar activity with reasonable errors ($\approx 7\%$), so they are very likely to be driven by the Sun. However, further investigation is required in order to identify if the periodicities with relative errors greater than 7% show influences from other physical or anthropogenic factors.

The time series analysis reveals the direct influence of the ongoing Popocatépetl eruption, which has being affecting its glaciers since 1999. On the other side, the Iztaccíhuatl and Citlaltépetl snow and ice coverage has a similar behaviour between them, departing from the behaviour shown by Popocatépetl volcano.

Future study should focus on:

1. Extending this approach to carry out snow and ice coverage forecasting with different artificial intelligence techniques such as neural networks.
2. The use of multispectral images, particularly those possessing a thermal/IR band in order to track data on the eruptive activity of Popocatépetl, for it to be included in future analyses.
3. The use of the snow and ice coverage time series obtained from the satellite images along with meteorological time series within the study region to generate a more robust pattern recognition and forecasting system for the region under study.

4. Including in the aforementioned pattern recognition system data about the geological history of the area in order to characterise it more precisely.

Acknowledgements Authors would like to thank G.E. Casillas-Aviña and A.D. Herrera-Ortiz for his support in pre-processing the initial satellite image dataset. This work was partially supported by projects SIP 20210925, 20211789, 20212034 and EDI grant, by Instituto Politécnico Nacional/Secretaría de Investigación y Posgrado, as well as by projects 8285.20-P and 10428.21-P from Tecnológico Nacional de México / IT de Mérida.

References

Almeida L, Cleef A, Herrera A, Velázquez A, Luna I (1994) El zacatonal alpino del Volcán Popocatépetl, México, y su posición en las montañas tropicales de América. Phytocoenologia 22(3):391–436

Almeida-Lenero L, Giménez de Azcárate J, Cleef A, Gonzales Trapaga A (2004) Las comunidades vegetales del zacatonal alpino de los Volcanes Popocatépetl y Nevado de Toluca. Región Central de México. Phytocoenologia 34(1):91–132

Al-Najjar HA, Kalantar B, Pradhan B, Saeidi V (2019) Conditioning factor determination for mapping and prediction of landslide susceptibility using machine learning algorithms. In: Earth resources and environmental remote sensing/GIS applications X, International Society for Optics and Photonics, vol 11156, p 111560K

Ames A, Hastenrath S (1996) Diagnosing the imbalance of Glaciar Santa Rosa, Cordillera Raura, Peru. J Glaciol 42(141):212–218

Andrade AO, Nasuto S, Kyberd P, Sweeney-Reed CM, Van Kanijn F (2006) EMG signal filtering based on empirical mode decomposition. Biomed Signal Process Control 1(1):44–55

Andrés N, Estremera DP, Zamorano JJ, Vázquez-Selem L (2010) Distribución del permafrost e intensidad de los procesos periglaciares en el estratovolcán Iztaccíhuatl (México) 1. Eria 83:291–310

Barth AM, Marcott SA, Licciardi JM, Shakun JD (2019) Deglacial thinning of the laurentide ice sheet in the adirondack mountains, New York, USA, revealed by 36cl exposure dating. Paleoceanogr Paleoclimatology 34(6):946–953. https://doi.org/10.1029/2018PA003477

Beaman JH (1962) The timberlines of Iztaccihuatl and Popocatepetl, Mexico. Ecology 43(3):377–385

Burroughs W (2005) Cycles and periodicities. In: Encyclopedia of world climatology, pp 173–177

Cong J, Gao C, Han D, Li Y, Wang G (2020) Stability of the permafrost peatlands carbon pool under climate change and wildfires during the last 150 years in the northern great Khingan mountains, china. Sci Total Environ 712. https://doi.org/10.1016/j.scitotenv.2019.136476

Cortés-Ramos J, Delgado-Granados H (2015) Reconstruction of glacier area on Citlaltépetl volcano, 1958 and implications for Mexico's deglaciation rates. Geofísica Int 54(2):111–125

De la Cruz-Reyna S, Siebe C (1997) The giant Popocatépetl stirs. Nature 388(6639):227–227

Delgado-Granados H (1996) Los glaciales del Popocatépetl: huéspedes efímeros de la montaña? Ciencias 041

Dikshit A, Pradhan B, Alamri AM (2020) Pathways and challenges of the application of artificial intelligence to geohazards modelling. Gondwana Research

Drakakis K (2008) Empirical mode decomposition of financial data. Int Math Forum 4:1191–1202

Ferrari L, Orozco-Esquivel T, Manea V, Manea M (2012) The dynamic history of the Trans-Mexican Volcanic Belt and the Mexico subduction zone. Tectonophysics 522:122–149

Gajek W, Trojanowski J, Malinowski M (2017) Automating long-term glacier dynamics monitoring using single-station seismological observations and fuzzy logic classification: a case study from Spitsbergen. J Glaciol 63(240):581–592

Ghil M, Vautard R (1991) Interdecadal oscillations and the warming trend in global temperature time series. Nature 350(6316):324–327

Gonzalez RC, Woods RE (2008) Digital image processing, 3rd edn. Pearson Education International

Heine K (1975) Permafrost am Pico de Orizaba/Mexiko. E&G Quat Sci J 26(1):212–217

Heine K (1994) Present and past geocryogenic processes in Mexico. Permafr Periglac Process 5(1):1–12

Hernandez-Capistran J, Martinez-Carballido J (2016) Thresholding methods review for microcalcifications segmentation on mammography images in obvious, subtle, and cluster categories. In: 2016 13th international conference on electrical engineering, computing science and automatic control, CCE 2016. https://doi.org/10.1109/ICEEE.2016.7751192

Hk Du, Jx Cao, Yj Xue, Xj Wang (2015) Seismic facies analysis based on self-organizing map and empirical mode decomposition. J Appl Geophys 112:52–61

Houssein E, El-din Helmy B, Oliva D, Elngar A, Shaban H (2021) Multi-level thresholding image segmentation based on nature-inspired optimization algorithms: a comprehensive review. Stud Comput Intell 967:239–265. https://doi.org/10.1007/978-3-030-70542-8_11

Huang NE (2001) Review of empirical mode decomposition. Wavelet Appl VIII Int Soc Opt Photonics 4391:71–81

Huang NE, Shen Z, Long SR, Wu MC, Shih HH, Zheng Q, Yen NC, Tung CC, Liu HH (1998) The empirical mode decomposition and the Hilbert spectrum for non-linear and non-stationary time series analysis. Proc R Soc Lond A: Math Phys Eng Sci 454(1971):903–995

Johnson C, Harrison C (1990) Neotectonics in central Mexico. Phys Earth Planet Inter 64(2–4):187–210

Kane R (2005) Short-term periodicities in solar indices. Solar Phys 227(1):155–175

Keller G (2008) Cretaceous climate, volcanism, impacts, and biotic effects. Cretac Res 29(5–6):754–771

Kim D, Oh HS (2009) EMD: a package for empirical mode decomposition and Hilbert spectrum. R J 1(1):40–46

Liu L, Cheng Y, Wang X (2017) Genetic algorithm optimized Taylor Kriging surrogate model for system reliability analysis of soil slopes. Landslides 14(2):535–546

Liu G, Zhao L, Li R, Wu T, Jiao K, Ping C (2017a) Permafrost warming in the context of step-wise climate change in the Tien Shan mountains, china. Permafr Periglac Process 28(1):130–139. https://doi.org/10.1002/ppp.1885

Lorenzo JL (1959) Los glaciares de México. Instituto de Geofísica, Universidad Nacional Autónoma de México, Technical report

Lugo-Hubp J, Robles-Padilla J, Eternod-Aguilar A, Ortuño-Ramírez V (1981) La disección del relieve en la porción centro oriental del Sistema Volcánico Transversal. Investigaciones geográficas 11:7–19

Macías JL (2007) Geology and eruptive history of some active volcanoes of México. Spec Papers Geoll Soc Am 422:183

Macías J, Arce J, García-Tenorio F, Layer P, Rueda H, Reyes-Agustin G, López-Pizaña F, Avellán D (2012) Geology and geochronology of tlaloc, telapón, iztaccíhuatl, and popocatépetl volcanoes, sierra nevada, central mexico. GSA Field Guides 25:163–193. https://doi.org/10.1130/2012.0025(08)

Mardia KV, Hainsworth T (1988) A spatial thresholding method for image segmentation. IEEE Trans Pattern Anal Mach Intell 10(6):919–927

Mendo-Pérez G, Arciniega-Ceballos A, Matoza R, Rosado-Fuentes A, Sanderson R, Chouet B (2021) Ground-coupled airwaves template match detection using broadband seismic records of explosive eruptions at Popocatépetl volcano, Mexico. J Volcanol Geotherm Res 419. https://doi.org/10.1016/j.jvolgeores.2021.107378

Mojica Moncada D, Cardenas C, Mojica J, Brondi F, Barragán Barrera D, Marangunic C, Holland D, Franco Herrera A, Casassa G (2020) The Lange Glacier and its impact due to temperature increase in the Admiralty Bay, King George Island, Antarctic Peninsula during the Austral Summer 2018–2019. AGU Fall Meet Abstr 2020:C062-0004

Moreno JL, Navarro F, Izaguirre E, Alonso E, Zabalza J, Revuelto J et al (2020) Glacier and climate evolution in the Pariacacá Mountains, Peru. Cuadernos de Investigación Geográfica 46(1):127–139

Nguvava M, Abiodun B, Otieno F (2019) Projecting drought characteristics over East African basins at specific global warming levels. Atmos Res 228:41–54. https://doi.org/10.1016/j.atmosres.2019.05.008

Nixon GT (1989) The geology of Iztaccíhuatl volcano and adjacent areas of the Sierra Nevada and Valley of Mexico, vol 219. Geological Society of America

Nunes JC, Bouaoune Y, Delechelle E, Niang O, Bunel P (2003) Image analysis by bidimensional empirical mode decomposition. Image Vis Comput 21(12):1019–1026

Orozco-del-Castillo MG, Hernández-Gómez JJ, Yañez-Casas GA, Moreno-Sabido MR, Couder-Castañeda C, Medina I, Novelo-Cruz R, Enciso-Aguilar MA (2019) Pattern recognition through empirical mode decomposition for temperature time series between 1986 and 2019 in Mexico City downtown for global warming assessment. In: International congress of telematics and computing. Springer, pp 45–60

Orozco-del-Castillo MG, Ortiz-Alemán JC, Couder-Castañeda C, Hernández-Gómez JJ, Solís-Santomé A (2017) High solar activity predictions through an artificial neural network. Int J Modern Phys C 28(06):1750075

Pare S, Kumar A, Singh G, Bajaj V (2020) Image segmentation using multilevel thresholding: a research review. Iran J Scie Technol Trans Electr Eng 44(1). https://doi.org/10.1007/s40998-019-00251-1

Pasquarè G, Vezzoli L, Zanchi A (1987) Morphological and structural model of Mexican Volcanic Belt. Geofísica Int 26(2)

Pérez-Moreno L, Rodríguez-Pérez Q, Zúñiga F, Horta-Rangel J, de la Luz Pérez-Rea M, Pérez-Lara M (2021) Site response evaluation in the Trans-Mexican Volcanic Belt based on HVSR from ambient noise and regional seismicity. Appl Sci (Switzerland) 11(13). https://doi.org/10.3390/app11136126

Pompeani DP, Bird BW, Wilson JJ, Gilhooly WP, Hillman AL, Finkenbinder MS, Abbott MB (2021) Severe little ice age drought in the midcontinental United States during the Mississippian abandonment of Cahokia. Sci Rep 11(1):1 8

Quincey D, Richardson S, Luckman A, Lucas RM, Reynolds J, Hambrey M, Glasser N (2007) Early recognition of glacial lake hazards in the Himalaya using remote sensing datasets. Global Planet Change 56(1–2):137–152

Rahman K, Maringanti C, Beniston M, Widmer F, Abbaspour K, Lehmann A (2013) Streamflow modeling in a highly managed mountainous glacier watershed using SWAT: the Upper Rhone River watershed case in Switzerland. Water Resourc Manag 27(2):323–339

Rangecroft S, Suggitt A, Anderson K, Harrison S (2016) Future climate warming and changes to mountain permafrost in the Bolivian Andes. Climatic Change 137(1-2):231–243. https://doi.org/10.1007/s10584-016-1655-8

Rodríguez-Pérez Q, Monterrubio-Velasco M, Zúñiga F, Valdés-González C, Arámbula-Mendoza R (2021) Spatial and temporal b-value characterization at Popocatépetl volcano, Central Mexico. J Volcanol Geotherm Res 417. https://doi.org/10.1016/j.jvolgeores.2021.107320

Rossotti A, Carrasco-Núñez G, Rosi M, Di Muro A (2006) Eruptive dynamics of the "Citlaltépetl Pumice" at Citlaltépetl volcano, Eastern Mexico. J Volcanol Geotherm Res 158(3-4):401–429. https://doi.org/10.1016/j.jvolgeores.2006.07.008

Russell S, Dewey D, Tegmark M (2015) Research priorities for robust and beneficial artificial intelligence. AI Mag 36(4):105–114

Siebe C, Macias JL, Abrams M, Rodriguez S, Castro R, Delgado H (1995) Quaternary explosive volcanism and pyroclastic deposits in east central mexico: implications for future hazards. In: Guidebook of geological excursions: in conjunction with the annual meeting of the geological society of America, New Orleans, Louisiana, November 6–9, 1995, pp 1–48

Song C, Sheng Y, Wang J, Ke L, Madson A, Nie Y (2017) Heterogeneous glacial lake changes and links of lake expansions to the rapid thinning of adjacent glacier termini in the Himalayas. Geomorphology 280:30–38. https://doi.org/10.1016/j.geomorph.2016.12.002

Soto-Molina VH, Delgado-Granados H (2020) Distribution and current status of permafrost in the highest volcano in North America: Citlaltepetl (Pico de Orizaba), Mexico. Geofísica Int 59(1):39–53. https://doi.org/10.22201/igeof.00167169p.2020.59.1.2079

Soto-Molina VH, Delgado-Granados H, Ontiveros-González G (2019) Estimación de la temperatura basal del "Glaciar Norte" del volcán Citlaltépetl, México. Modelo para determinar la presencia de permafrost subglaciar. Estudios Geográficos 80(287):e019–e019

Suarez-Gallareta E, Hernández-Gómez JJ, Cetzal-Balam G, Orozco-del-Castillo M, Moreno-Sabido M, Silva-Aguilera RA (2018) Sistema híbrido basado en redes neuronales artificiales y descomposición modal empírica para la evaluación de la interrelación entre la irradiancia solar total y el calentamiento global. Res Comput Sci 147(5):319–332

Terzi S, Torresan S, Schneiderbauer S, Critto A, Zebisch M, Marcomini A (2019) Multi-risk assessment in mountain regions: a review of modelling approaches for climate change adaptation. J Environ Manag 232:759–771

United States Geological Service (2021) Earthexplorer - home. https://earthexplorer.usgs.gov/. Accessed 23 Aug 2021

Veettil B, Wang S (2018) An update on recent glacier changes in Mexico using Sentinel-2A data. Geografiska Ann Ser A: Phys Geogr 100(3):307–318. https://doi.org/10.1080/04353676.2018.1478672

Velez ML, Euillades P, Caselli A, Blanco M, Díaz JM (2011) Deformation of Copahue volcano: inversion of InSAR data using a genetic algorithm. J Volcanol Geotherm Res 202(1–2):117–126

Villalpando O, Ik O (1968) Algunos aspectos ecológicos del volcán Nevado de Toluca. Master's thesis, Facultad de Ciencias, Universidad Nacional Autónoma de México

Wang H, Li Y, Liu Y, Huang G, Li Y, Jia Q (2021) Analyzing streamflow variation in the data-sparse mountainous regions: An integrated CCA-RF-FA framework. J Hydrol 596:126056

White SE (2002) Glaciers of México. In: Ferrigno J, Williams R (ed) (2002) Satellite image atlas of glaciers of the world, US Geological Survey, pp 383–405

White SE (1981) Neoglacial to recent glacier fluctuations on the volcano Popocatépetl, Mexico. J Glaciol 27(96):359–363

Yadollahie M (2019) The flood in Iran: a consequence of the global warming? Int J Occup Environ Med 10(2):54–56. https://doi.org/10.15171/ijoem.2019.1681

Yokohata T, Iwahana G, Sone T, Saito K, Ishizaki N, Kubo T, Oguma H, Uchida M (2021) Projections of surface air temperature required to sustain permafrost and importance of adaptation to climate change in the Daisetsu mountains, Japan. Sci Rep 11(1). https://doi.org/10.1038/s41598-021-94222-4

Zhang J, Li X, Yang R, Liu Q, Zhao L, Dou B (2017) An extended kriging method to interpolate near-surface soil moisture data measured by wireless sensor networks. Sensors 17(6):1390

Forest Degradation Estimation Through Trend Analysis of Annual Time Series NDVI, NDMI and NDFI (2010–2020) Using Landsat Images

Daniel Delgado-Moreno and Yan Gao(iD)

Abstract Forest degradation plays an important role in greenhouse gas (GHG) emissions and climate change. Previous research has shown that more GHG has been emit-ted through forest degradation than deforestation. Therefore, its monitoring and estimation is important for strategy design to combat climate change. In this work, we intend to estimate forest degradation in Ayuquila River Basin, Mexico through vegetation trend analysis using annual time series vegetation indices (2010–2020) specifically, NDVI (normalized difference vegetation index), NDMI (normalized difference moisture index), and NDFI (normalized difference fraction index) derived from Landsat images. The vegetation trend analysis was carried out using a linear regression model and tested by Mann-Kendall for significance. Slope coefficient was used to indicate the vegetation trend: positive slope indi-cates vegetation regrowth and negative slope indicates vegetation degradation. For forest degradation, only significant trends with negative slope were analyzed ($p < 0.05$). To discard negative trends due to deforestation, a forest mask was ap-plied both at the beginning and at the end of the analysis. The accuracy assessment showed that the forest degradation estimation by time series NDVI obtained the highest overall accuracy of 81.33%, followed by NDMI with 73.33% and fi-nally NDFI with 72%.

Keywords Forest Degradation · Time series analysis · Linear regression model · Spectral mixture analysis · NDVI · NDMI · NDFI

D. Delgado-Moreno (✉)
Tecnologias para la Informacion en Ciencias,Escuela Nacional de Estudios Superiores, Unidad Morelia, Universidad Nacional Autónoma de México, Morelia 58190, Mexico

Y. Gao
Centro de Investigaciones en Geografia Ambiental, Universidad Nacional Autónoma de México, Mexico City, Mexico
e-mail: ygao@ciga.unam.mx

R. Tapia-McClung et al. (eds.), *Advances in Geospatial Data Science*, Lecture Notes in Geoinformation and Cartography, https://doi.org/10.1007/978-3-030-98096-2_11

1   Introduction

With nearly four million hectares of area, forests cover about one third of the earth's surface (Blanc et al. 2016). Forests play a vital role in biodiversity conservation, regulating earth surface temperature and providing food and shelter. However, about two thirds of the forest areas are subject to anthropogenic activity: every year, millions of hectares of forest disappear due to deforestation and forest degradation (Anand et al. 2018). Both deforestation and forest degradation emit carbon, which account for 12–20 % of total anthropogenic greenhouse gas (GHG) emissions (Gregory 2005). However, the carbon emissions originating from forest degradation has been estimated from 40 to 212 % of those from deforestation (Baccini et al. 2017; Timothy et al. 2018). The big uncertainty is largely due to the difficulty in accurately mapping forest degradation.

Forest degradation is generally understood as disturbances in forest landscape causing changes in forest structure and canopy cover. The definitions vary depending on whether it is focused on ecosystem services or on carbon content (Food and Agriculture Organization 2011). For example, FAO defines forest degradation as a reduction of the capacity of the forest to provide goods and services (Food and Agriculture Organization 2010). From the climate change point of view, forest degradation is considered a process that results in forest carbon emissions but not in a change of land cover type (GOFI 2016). Remote sensing is essential in mapping the occurrence of forest disturbance and quantifying its extent and severity (Defries and Hansen 2000; Chloé 2020; Kenneth et al. 2016). Moderate to high resolution satellite sensors such as Landsat and Sentinel have been applied successfully in mapping disturbances from wildfires, swidden agriculture, and logging (Gregory 2005; Blanc et al. 2016; Loïc et al. 2016).

Time series analysis is a statistical method that can be used to determine patterns of behavior on a set of data collected over a specific period (Gilbert 1987). With the increasing number of earth observation data becoming available, new methods have been developed that use time series earth observation data for monitoring vegetation dynamics and mapping forest disturbances such as deforestation and forest degradation (Kenneth et al. 2016; Robert et al. 2010; Zhe et al. 2000–2014). Vegetation trend analysis is a simple but effective method to analyze vegetation dynamics including its greenness trends and its gradual changes (Zhe et al. 2000–2014; Yonatan and María 2020). Zhu et al. (2000–2014) analyzed vegetation greenness trends using all available Landsat 5, 7 and 8 images acquired between 1999 and 2014 and a slope coefficient. Tarazona et al. (2020) used NDVI (MODIS) to identify gradual changes in forest cover by applying a linear regression, and forest degradation was identified by adding significant negative trends (p value <0.05) from time series NDVI.

The main objective of this study was to estimate forest degradation through vegetation trends analysis using annual spectral indices including NDVI, NDMI, and NDFI derived from Landsat images (2010–2020). The temporal trends of the spectral indices were assessed using a slope coefficient tested by Mann-Kendall (MK) test, at a 95% confidence level. For forest degradation, only areas with significant negative vegetation trends were maintained.

2 Materials and Methods

2.1 The Study Area

The study area is the Ayuquila river basin in the western Pacific area of Mexico (Fig. 1). It includes municipalities of El Grullo, Autlán de Navarro, Unión de Tula, Tux-cacuesco, Tonaya, Zapotitlán de Vadillo, El Limón and Tolimán. The area has large topographic variations, with the elevation ranging from 260 m to 2500 m above mean sea level. The annual rainfall in average ranges from 800 mm to 1200 mm and the rain occurs mostly between June and October. The monthly temperature in average ranges from 18°C to 22°C (Cuevas 1998). There are both temperate forest and tropical dry forest in the study area. The temperate forest covers about 12 % of the watershed while tropical dry forest occupies around 24 % of the basin (Skutsch et al. 2012).

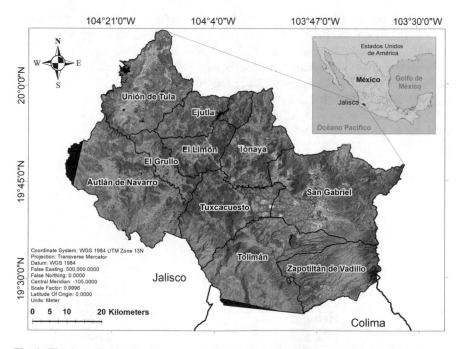

Fig. 1 The Ayuquila river basin represented by Landsat OLI 2020 image with natural color composite with bands combinations of R:6, G:5, B:4

2.2   Methods

2.2.1   Data Acquisition and Preprocessing

Annual Landsat Collection 1 data covering the study area for the period 2010–2020 was downloaded from Earth Explorer site of the United States Geological Survey. The collected data consists of Level-1 Terrain Precision Correction (L1TP) products which are radiometrically calibrated and orthorectified (USGS). Only images from the dry season were collected for monitoring degradation to minimize cloud cover Table 1. For atmospheric correction, the Dark Object Subtraction (DOS) algorithm was applied. The radiometric and atmospheric correction was carried out in R (R Core Team 2020) using the package RStoolbox.

2.2.2   Spectral Indices: NDVI and NDMI

Using the corrected Landsat images, NDVI was calculated by combining the near-infrared (NIR) and red (RED) bands, as shown in Eq. 1. The RED and NIR are the band 3 (0.626–0.693 µm) and band 4 (0.776–0.904 µm), respectively, of the Landsat TM and the band 4 (0.630–0.680 µm) and band 5 (0.845–0.885 µm), respectively, of the Landsat 8 OLI. NDMI was calculated using NIR and mid-infrared (SWIR) bands, as shown in Eq. 2. The NIR and SWIR are the band 4 (0.776–0.904 µm) and band 5 (1.55–1.75 µm), respectively, of the Landsat TM and the band 5 (0.851–0.879µm) and band 6 (1.566–1.651 µm), respectively, of the Landsat 8 OLI.

$$NDVI = \frac{NIR - RED}{NIR + RED} \tag{1}$$

$$NDMI = \frac{NIR - SWIR}{NIR + SWIR} \tag{2}$$

Table 1 Satellite image type and acquisition date

Year	Aquisition data	Path/Row	Satellite
2010	25/01/2021	29/46	Landsat-5 TM
2011	01/03/2011	29/46	Landsat-5 TM
2013	10/06/2013	29/46	Landsat-8 OLI
2014	25/03/2014	29/46	Landsat-8 OLI
2015	28/03/2015	29/46	Landsat-8 OLI
2016	30/03/2016	29/46	Landsat-8 OLI
2017	01/03/2017	29/46	Landsat-8 OLI
2018	20/03/2018	29/46	Landsat-8 OLI
2019	23/03/2019	29/46	Landsat-8 OLI
2020	25/03/2020	29/46	Landsat-8 OLI

2.2.3 Spectral Mixture Analysis and NDFI

NDFI was calculated using the fractions of green vegetation (GV), non-photosynthetic vegetation (NPV), soil (So) and shade (Sh), as shown in Eq. 4. The fractional images were derived using Spectral Mixture Analysis (SMA), an algorithm that allows the pixels to be decomposed into components. SMA assumes that the reflectance in a pixel (Ri) is the sum of the reflectances of each component within the pixel weighted by the respective proportional covers as shown in Eq. 3.

$$R_i = \sum_{j=1}^{Q} R_{ij} X_j + e_i \tag{3}$$

where R_{ij} is the reflectance of component j in band i, X_j is the area fraction of component j, e_i is the error term and Q is the number of components. In this study, we chose endmembers which represent the pure pixels for green vegetation (GV), non-photosynthetic vegetation (NPV), soil (So), and shadows (Sh). We applied the package RStoolbox in R for the SMA. Based on the results of the SMA, we calculated NDFI (Eq. 4).

$$NDFI = \frac{GV_{sh} - (NPV + S_o)}{GV_{sh} + (NPV + S_o)} \tag{4}$$

where GV_{sh} represents shadows of the green vegetation, and it is calculated following the Eq. 5.

$$GV_{sh} = \frac{GV}{100 - Sh} \tag{5}$$

2.2.4 Trend Analysis

The trend analysis was carried out using a simple linear regression model Eq. 6 on the time series NDVI, NDMI, and NDFI using function "lm" in R. The independent variable are the years, from 2010 to 2020, and the dependent variable are the spectral indices.

$$\widehat{Y} = a + bX + e \tag{6}$$

where $\widehat{Y}$ is the estimated value of the dependent variable, a is the intercept, b is the slope of the trend, X is the independent variable, and e is the error term. Slope b is calculated by Eq. 7.

$$b = \frac{\sum XY - n \sum \overline{XY}}{\sum X^2 - n \sum \overline{X^2}} \tag{7}$$

where X represents the values of the independent variable (year) and Y the values of independent variable (spectral index), and n is the number of data in each time series, in this case, 10.

2.2.5 Mann-Kendall Test

Mann-Kendal (MK) test is applied to check if there is a monotonic upward (increasing) or downward (decreasing) trend of the variable over time (Gilbert 1987; Mann 1945; Kendall 1975). The monotonic trend may be linear or non-linear. The MK was used to test if the slope of the linear regression line is different from zero. The MK test which is a non-parametric test that takes values from $+1$ to -1, where positive values indicate an increasing trend and negative values indicate a decreasing trend. Further-more, the higher the absolute value obtained by this test, the greater the congruence of the resulting trend (Priyanka 2020). The MK test analyzes each value by comparing it with the remaining values using Eq. 8, where n is the number of time series data.

$$S = \sum_{i=1}^{n-1} \sum_{j=i+1}^{n} sign(y_j - y_i) \tag{8}$$

When the comparison $(y_j - y_i)$ is positive value, it takes value of 1, and when it is negative it takes value of -1, and then it is 0, it takes value of 0. This S value provides information on the trend of the time series, if S is positive, indicates that the trend is increasing, when S is negative it indicates that there is a decreasing trend. The S value is necessary to obtain the value of the MK, which is calculated as Eq. 9:

$$t = S/(n(n-1)/2) \tag{9}$$

2.2.6 Accuracy Assessment

For accuracy assessment, random points were created for the two categories in the map: forest degradation and forest of no change. The total number of points n was determined following the good practices suggested by Olofsson (2014) as shown in Eq. 10.

$$n = \frac{(\sum W_i S_i)^2}{[S(\widehat{O})]^2 + (\frac{1}{N})W_i S_i^2} \approx (\frac{\sum W_i S_i}{S(\widehat{O})})^2 \tag{10}$$

where n is the number of pixels in the image, $\widehat{O}$ is the estimated standard error of the expected accuracy, which is usually set as 0.01. W_i is the area proportion of the class i and S_i is the standard deviation of i, which is calculated with the estimated user's accuracy of i, represented by U_i in Eq. 11.

$$S_i = \sqrt{U_i(1 - U_i)} \qquad (11)$$

The calculated total number of random points are assigned to each of the class proportional to the area of each class.

The overall accuracy, and user's and producer's accuracy were calculated through a confusion matrix (Congalton and Green 2019).

3 Results

3.1 Vegetation Trend Analysis with Linear Regression

The result of the estimated forest degradation by time series spectral indices is presented in (Fig. 2). The areas of forest degradation were identified as areas with negative slope of the trend that are significant ($P < 0.05$). To exclude deforested areas, forest mask was applied to make sure the detected degraded areas were both in forest category at the beginning (2010: over accuracy 88 %) and the end (2020: overall accuracy 89 %) of the analysis.

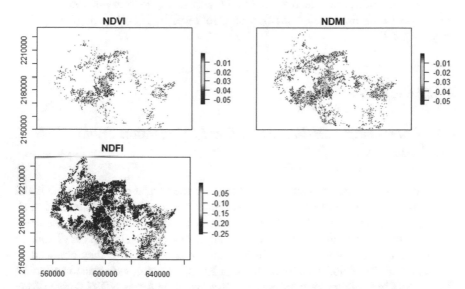

Fig. 2 The results of vegetation trend analysis with time series analysis of spectral indices of NDVI, NDMI, and NDFI. The legend presents the negative values of slope which were tested significant ($p < 0.05$) to indicate possible forest degradation

Table 2 The overall accuracy, user's accuracy and producer's accuracy of the estimated degraded forest by each of the spectral indices

Spectral index	Overall accuracy%	User's accuracy%		Producer's accuracy %	
		Degradation	No change	Degradation	No change
NDVI	81.33	85.33	77.33	79.01	84.05
NDMI	73.33	62.66	84.00	79.66	69.23
NDFI	72.00	52.00	92.00	86.66	65.71

3.2   Results of Accuracy Assessment

To determine which index had better results, the accuracy of the identified de-graded areas was evaluated. The number of random points was calculated as Eq. 12.

$$n \approx \left(\frac{\sum W_i S_i)}{S(\widehat{O})} \right)^2 = \left(\frac{13 * 0.950 + 678 * 0.985}{0.01} \right)^2 \tag{12}$$

After obtaining the random points, estimated forest degradation was evaluated and the results obtained are presented in Table 2. Forest degradation estimated by time series $NDVI$ obtained the highest accuracy, followed by $NDMI$, and then by time series $NDFI$.

4   Discussion

4.1   Comparing with Similar Studies in Vegetation Trend Analysis

Comparing with some other studies that have also used linear regression models and time series analysis to identify vegetation trends and forest degradation (Zhe et al. 2000–2014; Yonatan and María 2020), our study obtained slightly lower accuracy. For example (Zhe et al. 2000–2014) applied time series NDVI and EVI indices to analyze forest cover loss and gain in Guangzhou, China, and they obtained an overall reliability of 90.31 %. In the study of Kendall (1975), which analyzed forest degradation in southern Pero "Madre de Dios" by time series NDVI derived from MODIS sensor. They found 2.4 % of their study area presented degradation. However, they didn't report accuracy of the obtained results.

We also compared our results with the Global Forest Watch (GFW), which is an online platform that has updated data of forest cover change in various parts of the world (https://data.globalforestwatch.org/maps/tree-cover-loss-1/explore). Figure 3 shows the areas identified as forest degradation by NDVI time series analysis (black

Fig. 3 Forest degradation by vegetation trend analysis and forest loss estimated by Global Forest Watch

color) contrasted with forest loss data by GFW in pink color. Forest degradation by time series NDVI shows a clustered pattern (Group A and B), and only group B seems to be located in the same areas identified by GFW. Since GFW only presents forest loss data, it might miss subtle changes in forest cover as identified by trend analysis.

4.2 Spectral Indices of NDVI, NDWI and NDFI

Among the tested spectral indices, NDVI and NDMI are readily comparable since they depict vegetation greenness and moisture content which can be used as an indication of vegetation health (Bullock et al. 2020). The higher the values in NDVI and NDMI, the better condition the vegetations have. NDFI is a fraction index derived from SMA. The high NDFI value indicates intact forest, while fragmented and degraded forests have low NDFI values. In this study, we did not achieve expected results through NDFI, the possible reasons include the difficulty in locating the "pure" endmembers from Landsat images. In this study, we tested the significance of the slope coefficient by MK test. Other ways to refine the trend analysis is to apply a threshold of slope coefficient using expert knowledge and/or through field observation.

Acknowledgements This work was funded by the Consejo Nacional de Ciencia y Tecnología (CONACYT) 'Ciencia básica' SEP-285349 "Análisis del patrón espacial de la degradación en selvas y bosques de México con percepción remota en múltiples escalas en el tiempo y espacio". The authors are grateful of the help from Hind Taud in compiling the PDF file in Latex.

References

Anand A, Singh SK, Kanga S (2018) Estimating the change in forest cover density and predicting NDVI for west Singhbhum using linear regression. Int J Environ Rehabil Conserv 9:193–203

Asner GP, Knapp DE, Broadbent EN, Oliveira PJC, Keller M, Silva JN (2005) Selective logging in the Brazilian amazon. Science 310:480–482

Baccini A, Walker W, Carvalho L, Farina M, Sulla-Menashe D, Houghton RA (2017) Tropical forests are a net carbon source based on aboveground measurements of gain and loss. Science 358(6360):230–234

Blanc L, Gond V, Minh DH (2016) Remote sensing and measuring deforestation

Bullock EL, Woodcock CE, Olofsson P (2020) Monitoring tropical forest degradation using spectral unmixing and Landsat time series analysis. Remote Sens Environ 238:110968

Congalton K, Green RG (2019) Assessing the accuracy of remotely sensed data - principles and practices, 3rd edn. CRC Press, Boca Raton, USA

Cuevas RNNMGFSM (1998) El bosque tropical caducifolio en la reserva de la biosferasierra man-antlan, jalisco-colima, méxico. Bol, IBUG

Defries RS, Hansen MC (2000) Global continuous fields of vegetation characteristics: a linear mixture model applied to multi-year 8 km AVHRR data. Int J Remote Sens 21(6–7):1389–1414

Dupuis Chloé, Lejeune Philippe, Michez Adrien, Fayolle Adeline (2020) How can remote sensing help monitor tropical moist forest degradation? A systematic review. Remote Sens 12:1087

Dutrieux LP, Jakovac CC, Latifah SH, Kooistra L (2016) Reconstructing land use history from Landsat time-series. case study of a swidden agriculture system in Brazil. Int J Appl Earth Obs Geoinf 47:112–124

Food and Agriculture Organization (2010) Global forest resources assessment. FAO

Food and Agriculture Organization (2011) Assessing forest degradation, towards the development of globally applicable guidelines. FAO

Gilbert RO (1987) Statistical methods for environmental pollution monitoring

GOFI (2016) Integration of remote-sensing and ground-based observations for estimation of emissions and removals of greenhouse gases in forests: methods and guidance. Global forest observation initiative

Grogan Kenneth, Pflugmacher Dirk, Hostert Patrick, Verbesselt Jan, Fensholt Rasmus (2016) Mapping clearances in tropical dry forests using breakpoints, trend, and seasonal components from MODIS time series: Does forest type matter? Remote Sens 8:657

Kendall MG (1975) Rank correlation methods

Kennedy RE, Yang Z, Cohen WB (2010) Detecting trends in forest disturbance and recovery using yearly landsat time series: 1. landtrendr - temporal segmentation algorithms. Remote Sens Environ 114:2897–2910

Mann HB (1945) Non-parametric tests against trend, econometrica

Olofsson P (2014) Good practices for estimating area and assessing accuracy of land change. Remote Sens Environ

Pearson TR, Bernal B, Hagen SC, Walker SM, Melendy LK, Delgado G (2018) Remote assessment of extracted volumes and greenhouse gases from tropical timber harvest. Res Lett 13:065010

Priyanka JFV (2020) "freygeospatial,". https://freygeospatial.github.io/PM25-TimeSeries-R-Tutorial/. Accessed 18 Mayo 2021

Skutsch M, Martinez R, Morfin J, Allende T, Vega E, Morales J, Ghilardi A, Jardel E (2012) Analisis de cambio de cobertura y uso del suelo, escenario de referencia de car-bono y diseño preliminar del mecanismo de monitoreo, reporte y verificacion en los diez municipios de la junta intermunicipal de rio ayuquila [land cover and land use change analysis, reference scenario of carbon and preliminary design of the monitoring, reporting and verification system in the ten municipals of rio ayuquila], jalisco. Report

Tarazona Y, Miyasiro-López M (2020) Monitoring tropical forest degradation using remote sensing. challenges and opportunities in the Madre de Dios region, Peru. Remote Sens Appl Soc Environ 19:100337

Team RC (2020) A language and environment for statistical computing. R foundation for statistical computing

Zhu Zhe, Yingchun Fu, Woodcock Curtis E, Olofsson Pontus, Vogelmann James E, Holden Christopher, Wang Min, Dai Shu, Yang Yu (2016) Including land cover change in analysis of greenness trends using all available Landsat 5, 7, and 8 images: a case study from Guangzhou, china (2000–2014). Remote Sens Environ 185:243–257

Geospatial Data and Pandemic-Related Problems and Solutions

Estimating Importation Risk of COVID-19 in Hurricane Evacuations: A Prediction Framework Applied to Hurricane Laura in Texas

Michelle Audirac, Mauricio Tec, Enrique García-Tejeda, and Spencer Fox

Abstract In August 2020 as Texas was coming down from a summer COVID-19 surge, forecasts suggested that Hurricane Laura was tracking towards 6M residents along the East Texas coastline threatening to spread COVID-19 across the state. To assist local authorities facing the dual threat, we developed a framework that integrates evacuation dynamics and local pandemic conditions to quantify COVID-19 importations due to hurricane evacuations. For Hurricane Laura, we estimate that 499,500 [90% Credible Interval (CI): 347,500, 624,000] people evacuated the Texan counties, and that there were 2,900 [90% CI: 1,700, 5,800] importations of COVID-19 across the state. To demonstrate the transferability of the framework, we apply it to a scenario with characteristics matching those of Hurricane Rita, where a much feared direct hit towards the highly populated Houston/Galveston area was forecasted. For this scenario we estimate 1,054,500 evacuations [90% CI: 832,500, 1,162,000], and 6,850 COVID-19 importations [90% CI: 4,100, 13,670]. Overall, we present a flexible and transferable framework that captures spatial heterogeneity and incorporates geographic components for anticipating potential epidemiological risks resulting from evacuation movement due to hurricane events.

M. Audirac (✉) · M. Tec · S. Fox
The University of Texas at Austin, Austin, TX, USA
e-mail: michelle.audirac@austin.utexas.edu

M. Tec
e-mail: mauriciogtec@utexas.edu

S. Fox
e-mail: fox@austin.utexas.edu

E. García-Tejeda
Centro de Investigación y Docencia Económicas, Mexico City, Mexico
e-mail: enrique.garciatejeda@cide.edu

© The Author(s), under exclusive license to Springer Nature Switzerland AG 2022
R. Tapia-McClung et al. (eds.), *Advances in Geospatial Data Science*, Lecture Notes in Geoinformation and Cartography, https://doi.org/10.1007/978-3-030-98096-2_12

1   Introduction

Identifying which populations travel and how they spread in an evacuation is essential to assess the dual threat of a hurricane and an ongoing epidemic, yet anticipating hurricane evacuation is challenging. The relation between spatial patterns of storm surges, the geographic distribution of potential destinations, and the decisions that a household makes when threatened by an approaching hurricane is a highly complex and dynamic process (Baker 1991). Our aim is to offer the ability to anticipate individuals' movement in an evacuation before a hurricane hits land in the presence of an already prevalent regional epidemic, and quantify the importation and exportation of cases in both the origin and destination counties.

Geospatial Modeling of Storm Surges and other GIS tools have become an important tool in assessing the risk of hurricanes (Ferreira et al. 2014). For this work, we examine existing models that combine geospatial features, storm surge risk, and statistical methods to integrate a modeling framework that, along with prevalence numbers of an infectious disease, can estimate importation risk during a hurricane evacuation. We apply our framework to the evacuation in the Texas region in response to the approach of Hurricane Laura during the COVID-19 pandemic. While the results presented here pertain to these specific geographic features, demographic characteristics, hurricane, and epidemic settings, the framework itself is flexible for application in future hurricane events and other geographical locations undergoing an epidemic.

The remainder of this study is organized as follows. First, we provide a general description of the events surrounding Hurricane Laura—including the prevalence of COVID-19 cases in Texas at the time of the approach of the hurricane and an account of which counties issued mandatory or voluntary evacuation orders. We examine the hurricane literature and provide a background that explains the choice of models that integrate our proposed framework. The methodology description is presented next, followed by the results of the framework implementation in our case study. To highlight the applicability of the framework to other scenarios, a counterfactual scenario is explored with corresponding conditions to those experienced for Hurricane Rita in 2005 under the assumption that an ongoing epidemic was present at the time. For validation purposes, comparisons of point estimates with reported evacuation and reception numbers is presented.

2   Case Scenario and Study Area

Amid the COVID-19 pandemic, Hurricane Laura approached eastern Texas coastline and was predicted to the make landfall on August 27 2020. To identify counties threatened by the forecast track of Laura, we used news coverage and State Situation Reports. A Hurricane Warning is typically declared in a region when the onset of hurricane conditions is likely within 24 h. Figure 1 shows the projected trajectory of Laura the day before the predicted landfall and the layout of counties that

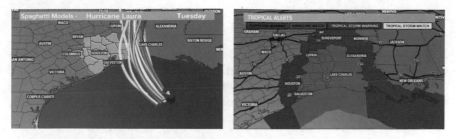

Fig. 1 The projected trajectory of Laura the day before the predicted landfall and the layout of counties that issued a hurricane warning (KHOU News 2020)

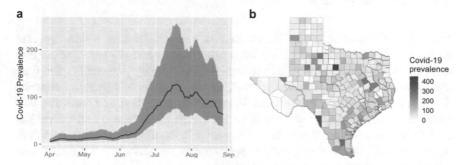

Fig. 2 Estimated COVID-19 prevalence per 10,000 people in Texas: **a** since April 2020, around the first identified COVID-19 case in the US. **b** spatial distribution on the week prior to Hurricane Laura's landfall on August 2020

issued a hurricane warning. Within the highlighted region, 11 Texan counties issued mandatory evacuation orders, and 3 issued voluntary orders.

During the days preceding Hurricane Laura's landfall and all through August, the second surge of cases in Texas appeared to be mostly under control. COVID-19 cases in both the coastal counties and inland Texas had a downwards trend. However, since the reported number of cases do not necessarily reflect the absolute number of people infected, we used the method by Fox et al. (2021) to estimate a lower, median and upper bound of disease prevalence. The incidence estimation assumes that confirmed COVID-19 case counts represent somewhere between one third and one tenth of all infections. Figure 2a shows the COVID-19 prevalence per 10,000 residents in Texas from the beginning of the pandemic to the end of September, and panel Fig. 2b displays its spatial distribution estimated for the week of Aug 20–27. Overall, the peak of estimated cases is around 100 infections per 10,000 in mid August, with a substantial upper bound of up to 200 infections. The figure also shows that at that time, counties with higher prevalence were present in some inland regions, and not coastal areas.

3 Background

In this section we examine existing models in the hurricane literature and specifically consider those that incorporate spatial variables and capture geographic and demographic heterogeneity. Emphasis is put on the methods' ability to be applied over regions with varying geographic attributes, population distributions, as well as different hurricane strengths and levels of disease prevalence.

3.1 Evacuation Rates

Methods that focus on estimating evacuation rates fall mainly within two categories (Wilmot and Mei 2004): (1) household models and (2) participation rates approaches. In the case of household models, the primary purpose is identifying associations between household attributes and the decision of whether to evacuate. These attributes include demographic features, risk perception, and hurricane characteristics. These studies provide valuable knowledge about human decision-making under a hurricane emergency situation. However, most of the household models do not apply to an approaching hurricane event, nor do they take into account the geographic characteristics of the threatened regions and varying levels of surge or storm risk.

In the participation rates approaches, geographic areas are assigned estimates of evacuation rates based on key features such as hurricane strength and surge probability. The Federal Emergency Management Agency (FEMA) owns a comprehensive proprietary solution called Evacuation Traffic Information System (ETIS) (Palmer 2001) that, along hurricane traffic simulations, computes evacuation rates. The applicability of the participation rate approaches make them more suitable than household models for a framework whose emphasis is on preparation for specific storm. Nonetheless, access to previously trained participation rates methods is not publicly available.

3.2 Destinations

Two decisions lead to choosing a particular evacuation target location: (1) the accommodation type choice; and (2) the destination choice. An accommodation type is a sheltering option. Common alternatives include houses of friends and relatives, hotels, and public shelters (Mesa-Arango et al. 2013). A destination choice, in constrast, is concerned with the flow of evacuees among predefined evacuation regions, or with the zone distribution of their accomodation.

Accomodation type choice Mesa-Arango et al. (2013) highlight the lack of evidence and prior rigorous statistical models explaining accommodation type demand. And, as far as we are aware, their work is the only one providing a statistical approach to model the accomodation type choice. Instead, the proportion of evacuees traveling to each accommodation type is typically a fixed proportion informed from a specific past hurricane.

In terms of applicability, Mesa-Arango's model (2013) is not appropriate for prediction in emergency situations because several of its input variables, such as previous experience with hurricanes, or being required to work during the evacuation, depend on targeted survey responses.

Destination choice Among studies modeling the destination choice, three types of models can be discerned: the gravity model (Modali 2005), intervening opportunity model (Wilmot and Mei 2004), and multinomial logit model (Cheng et al. 2008). These studies estimate zonal distribution of hurricane evacuation trips using a telephone survey conducted in South Carolina following Hurricane Floyd (Dow and Cutter 2002), thereby obtaining two origin-destination (OD) matrices: one for evacuees going to friends and family; and the other for those staying in hotels. Both the gravity and the intervening opportunity models learn the influence of some measure of travel impedance, such as a function of travel time and distance, on the distribution of origin-destination trips. In the multinomial logit model, the probability of choosing a destination from a given location is estimated by incorporating demographic characteristics of destinations.

4 Modeling Framework

We package methods into modules and build a framework that consists of two modules aimed at the following objectives:

1. The estimation of evacuation rates per coastal county.
2. The prediction of destination choices to obtain expected receptions in inland counties.

Local prevalence estimates of COVID-19 are combined with evacuations and reception numbers from the framework's modules to determine which counties are likely to face the highest importation risk.

4.1 Evacuation Rates Module

The first module of the framework consists of a participation rates approach that determines evacuation rates based on hurricane intensities and evacuation zone categories—which are geographic areas distinguished by their vulnerability to hurricanes of different strengths. Even though there is no convention on how jurisdictions determine hurricane evacuation zones, many jurisdictions use National Oceanic and

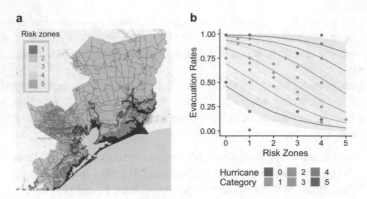

Fig. 3 **a** Coastal zones are assigned a risk level using NOAA's SLOSH model (Jelesnianski 1992). A risk zone delineates the near worst-case scenario of flooding for each hurricane category. Hurricanes categories are in order of increasing intensity. A risk area 3 has the potential of flooding with hurricanes of category 3, 4 and 5. **b** Regression results showing an inverse relationship between risk zone and evacuation rates and direct relationship between hurricane category and evacuation rates

Atmospheric Administration (NOAA)'s SLOSH surge model (Jelesnianski 1992) (see Fig. 3a) as we do for our analysis.

The proposed model is a weighted generalized linear regression, informed by both evacuation expectations of coastal residents and observed compliance of evacuation orders in the region. The dependent variable considered is the probability of evacuation rate, modeled as a conditional Beta distribution, while the predictors are the risk zones and the hurricane categories. The reader should refer to Sect. 7 in the Appendix for additional model details.

To obtain data for estimating the parameters in the model, we compiled a database from multiple previous studies on Hurricane evacuations in the Gulf of Mexico area. These studies are Kang et al. (2007), Lindell et al. (2005), Morrow and Gladwin (2005), Knabb et al. (2006), Huang et al. (2012) and Huang et al. (2012). The resulting database—available with the code repository accompanying this paper at https:// github.com/audiracmichelle/hurricane_covid—contains 45 entries.

The predicted conditional evacuation rates estimated from the model are shown in Fig. 3b. The resulting values confirm the intuition that the highest rate occurs in those zones that are most vulnerable and that rates increase with hurricane intensities.

Regions under mandatory evacuation orders are assigned an estimate of total number of evacuees by multiplying the population in each evacuation zone by the evacuation rate corresponding to their surge risk and hurricane category. Regions under voluntary evacuation orders, are classified as risk zones for Hurricane category 0. The number of exportations is roughly the number of evacuees times the disease prevalence in a given area.

Table 1 Variables used in the origin-destination model

Friends/relatives	Hotel
Miles from origin to destination	Miles from origin to destination
Destination population	Destination number of hotels
Hurricane threatened area (binary variable)	Hurricane threatened area (binary variable)
Metropolitan (MSA) indicator	Interstate highway indicator
Percentage white	Percentage white

4.2 Destinations Module

For this module we adopt the model of Cheng et al. (2008) whose multinomial logit regression uses a set of attributes that influence the preference for a destination alternative according to the accommodation type as listed in Table 1.

The values of the covariates for our case study are obtained from three main sources: (1) geographic features from the R package geosphere (Hijmans 2019); (2) demographic characteristics from the American Community Survey (U.S. Census Bureau 2018); evacuation routes in Texas from the Homeland Infrastructure Foundation-Level Data API (Homeland Infrastructure Foundation-Level Data 2020), and hurricane forecast tracks from the National Hurricane Center GIS products (National Hurricane Center Data in GIS Formats 2020). The values for the regression coefficients of these predictors are taken directly from the estimates in Cheng et al. (2008).

Following Kang et al. (2007), we established fixed weights for the preference for accommodation type, with 60% of evacuees going to family and friends and the remaining 40% staying in hotels. Kang et al. (2007) derived these proportions based on survey responses. The model output consists of two origin-destination (OD) matrices, one for each accommodation type. Using the exportation numbers from the evacuation rates module and combining them with the estimated OD matrices gives the expected number of total importations for each evacuation destination choice. Section 8 in the Appendix presents additional details on the statistical model.

5 Results

The proposed framework is applied to the scenario where Hurricane Laura is forecasted to make landfall within the next 24 h. In this timeframe, Laura was expected to reach the shore with a category-4 strength with a total of 6,016,750 residing in the counties under a hurricane warning. Under this scenario, a total of 463,473 and 707,224 residents living in the surge risk zones are assumed to be under mandatory and voluntary orders to evacuate.

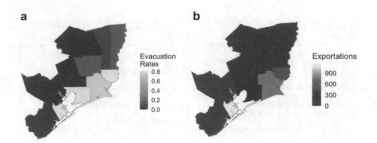

Fig. 4 **a** County-level evacuation rate prediction. Evacuation rates for counties were computed as the weighted averages of the census tract rates, where the weight for each tract was the percentage of the region population it contains. **b** Estimated number of total exported infections by county. Disease prevalence is estimated assuming 20% infection detection rate. Evacuation rates and prevalence jointly determine the number of total exportations

We obtain estimates of evacuation rates for census block groups (CBG) in the region as described in the Methodology section. The median predicted evacuation rate of CBG's in risk zones with mandatory evacuation orders is 85%, with 90% the distribution of CBG predictions ranging from 69% to 95%. In contrast, CBG's risk zones with voluntary orders had a median predicted evacuation rate of 6%, and ranged between 3 and 19%. After aggregating the CBG results in counties we find that Orange County has the highest estimated evacuation rate, which is predicted to be 80%.

Multiplying the population in each county by the appropriate evacuation rate and disease prevalence provides an estimate of the total number of exportations. We estimate that 499,500 [90% CI: 347,500, 624,000] people evacuated the Texan counties that issued an evacuation order. With prevalence in the region revolving around 66 per 10,000 people, the total number of predicted COVID infection exportations is 2,900 [90% CI: 1,700, 5,800]. Figure 4 contains the spatial distribution of the county-level evacuation rates and exportation's estimates.

For each accommodation type, assuming 60% of evacuees go to hotels and 40% to family/relatives, we obtain a mapping of probabilities for all origin and destinations counties using the multinomial logit model described in the Methodology section. These probabilities along with the predictions for the number of evacuees captured by the evacuation rates module produce estimates of receptions for the destination counties.

Results show that travels disperse outside the threatened region; thus, no single county accumulates more than 2.5% of the evacuees. When outcomes are aggregated in District level, the number of receptions (and receptions relative to the population of the district) in major districts include: San Antonio 44,800 (1.7%), Austin 45,500 (1.9%), Dallas 32,500 (0.65%), and Fort Worth 29,000 (1.1%). Relative to its population, reception of evacuees for counties in Yoakum District (Austin, Calhoun, Colorado, DeWitt, Fayette, Gonzales, Jackson, Lavaca, Matagorda, Victoria and Wharton) is of 17%, which is the highest expected value in the case scenario.

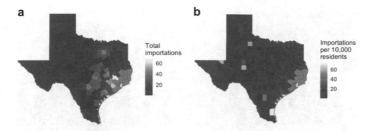

Fig. 5 Importation risk distribution: **a** Estimated imported infections per destination county. **b** Importations per 10,000 residents. In general evacuees were likely to move to regions with higher population densities. When aggregated in Districts, importations represent at most 10 infected people per 10,000 residents not exceeding 2 per 10,000 residents in the most populous Districts

Analogous results of imported infections show that county importations range between 3 and 18 (IQR); and no county imports more than 68 infections. Importation estimates (and importation per 10,000 residents) aggregated in Districts are as follows: San Antonio 257 (0.98), Austin 261 (1.11), Dallas 189 (0.35), and Fort Worth 169 (0.64). In Yoakum importations are estimated to be 333, corresponding to 10 importations per 10,000 residents. The dispersion of importations in the destination counties appears in Fig. 5.

To validate and assess the transferability of the framework, we simulate a counterfactual scenario assuming conditions similar to those of Hurricane Rita whose feared forecast track with a category-5 strength pointing directly towards the highly populous area of Houston/Galveston. Counties in the forecast tracks of Hurricane Rita and Hurricane Laura appear in Fig. 6. The results for the evacuation rates module applied to the counterfactual scenario yield 1,054,500 expected evacuations [90% CI: 832,500, 1,162,000] and 6,850 exportations [90% CI: 4,100, 13,670].

We compare these results with reported evacuations for Rita. According to a media survey (Jeff Lindner Report on Hurricane Rita 2020), 2.5 million people evacuated, 59.9% of which were *not* residents in an evacuation risk zone. Therefore, approximately 1.0025 M (2.5 ∗ (1 − 0.599) came from risk zones. Adjusting by a 20% population growth (1.0025 ∗ 1.20 = 1.203)), this value is 15% above our 1.054 estimate. We also compare destinations. Of those respondents who evacuated, 32% headed for Austin, San Antonio or Dallas/Fort Worth. Our estimated receptions for those cities is 323,000, which represents 30% of the total receptions. Overall, these results show that the model's predictions are close to reported data, adding evidence that the model is a useful prediction tool for evaluating alternative scenarios.

Another point validation was possible with evacuee data provided by the City of Austin Homeland Security and Emergency Management, which reported 45,000–50,000 evacuees in Austin City due to Hurricane Laura. This range coincides with our receptions estimate.

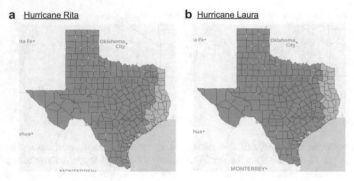

a Hurricane Rita   **b** Hurricane Laura

Fig. 6 Counties in the path of the forecast tracks of both Hurricane Laura (2020) and Hurricane Rita (2005), representing the case study and counterfactual scenarios of our study

6 Discussion

This study presents a framework that allows anticipating the impact of hurricane evacuations in an ongoing epidemic. The modules of the framework focus on predicting the size and dispersion of moving populations due to an approaching hurricane. Paired with prevalence numbers, these predictions produce estimates of disease exportation and importation numbers.

We underscore the framework's ability to capture spatial heterogeneity and incorporate geographic components that influence hurricane evacuation processes. Indeed, surge risk zones and preferred destinations are not homogeneously distributed geographically. Therefore, in the evacuation rates module, zones with higher surge risk are associated with higher participation rates. Similarly, in the destinations module, distance to highways and concentration of hotels are attributes that affect the preference for a certain destination in addition to distance between origin and destination counties. Thus, even if rigorous validation of the models' estimates remain to be performed, the point validations we present provide assurance that the models that integrate the framework generate likely evacuation patterns relative to the path of an approaching storm.

This study suggests few new directions for future research. First, an important limitation in the current framework is the omission of shadow evacuations which are voluntary displacement of people from areas outside a declared evacuation area. These ideally should not occur as they can congest roadways unnecessarily. Including shadow evacuations in the framework presents several challenges. For instance, the magnitude of the shadow evacuation for Hurricane Rita exceeded expectations (Jeff Lindner Report on Hurricane Rita 2020). There is general agreement that the severity of Hurricane Katrina which preceded Rita by less than a month and widespread media coverage (Jeff Lindner Report on Hurricane Rita 2020) had a strong influence over Rita's shadow evacuees. Therefore, behavior change influenced by past recent events might dramatically affect evacuation predictions. Second, answering the question of whether the presence of an ongoing epidemic modifies evacuation decisions is out of the scope of this work. Nonetheless, it might be the case that instead of going to friends and family, evacuees decide to go to hotels to avoid spreading a disease. Finally, exportation and importation cases alone do not explain the transmission impact of evacuations. Importations might have a direct connection with disease spread, regardless ubiquitous behavior and overall population connectedness are the key determinants of new surges. In addition, mask usage and social distancing can limit the impact of temporal relocation of evacuees. Future work should be concerned with the interplay of these factors to determine transmission impact due to importations.

Our framework's flexibility and applicability is useful for disaster preparedness in the co-occurrence of hurricanes and epidemics. Combining the geographic characteristics' of the evacuation region with the overall magnitude of the evacuation and the specific characteristics of an epidemic can help authorities and emergency officials understand the importation risk that may accompany a hurricane and guide emergency planning accordingly.

7 Appendix: Statistical Model for Predicting Evacuation Rates

This section presents a simple weighted generalized linear model for predicting evacuation rates given a combination of risk zones and Hurricane categories.

We first introduce some notation. Let us denote y_i the registered evacuation rate for all $i = 1, \ldots, N$. Similarly, let $z_i \in \{0, \ldots, 5\}$ be the indicator of risk zone for and $h_i \in \{0, \ldots, 5\}$ the indicator of hurricane intensity.

Then, the statistical model is described as the following generalized Beta regression parameterized by a mean and precision parameter

$$y_i \sim \text{Beta}(\mu_i \lambda, (1 - \mu_i)\lambda)$$
$$\text{logit}(\mu_i) = \alpha + \beta_{z_i}^{\text{zone}} + \beta_{h_i}^{\text{intensity}}.$$

where $\theta := (\alpha, \{\beta_j^{\text{zone}}\}_{j=0}^5, \{\beta_k^{\text{intensity}}\}_{k=0}^5)$ are learnable parameters. Using this parameterization $\mathbb{E}[y_i] = \mu_i$ and $\text{Var}[y_i] = \mu_i(1 - \mu_i)/(\lambda + 1)$.

Observations in the complied training dataset (see Sect. 4.1) fall mainly within two categories: first, 30 instances where the reported rate is the actual observed evacuation rate from a retrospective study; second, 15 cases where the recorded rate is based on reported intentions to evacuate based on survey responses.

Finally, we note that observations carry different weights depending on whether or not the evacuation rate is observed or intended. For this purpose, we let ω_i be a confidence weight such that $\omega_i = 1$ if y_i is the report from actual observed evacuation rate, and $\omega_i = 1/2$ if the data source is a reported intention to evacuate. Then θ is obtained by a weighted maximum likelihood with the weights defined by $\{\omega_i\}_{i=1}^N$.

8   Appendix: Multinomial Logit Model for Origin-Destination Prediction

Let use denote by P_{ij} the probability of choosing destination i given origin j. The model takes the following form:

$$P_{ij} = \frac{\exp(\beta^\top x_{ij})}{\sum_{k=1}^{K} \exp(\beta^\top x_{ik})}$$

where the x_{ik}'s are the vectors of geographic, demographic and infrastructure features representing destination k and the cost of traveling from i to k for $k = 1, \ldots, K$ alternatives. These features are listed in Table 1.

Acknowledgements   We thank Dr. Kelly Gaither and Dr. Gordon Wells of the University of Texas at Austin, as well as Mario Chapa from the Texas Division of Emergency Management, whose insights helped us better design and inform our models.

References

Baker EJ (1991) Hurricane evacuation behavior. Int J Mass Emerg Disast 9(2):287–310
Castro LA, Fox SJ, Chen X, Liu K, Bellan SE, Dimitrov NB, Galvani AP, Meyers LA (2017) Assessing real-time Zika risk in the United States. BMC Infect Dis 17(1):284
U.S. Census Bureau (2018) American community survey. https://www.census.gov/programs-surveys/acs/news/data-releases.html
Cheng G, Wilmot CG, Baker EJ (2008) A destination choice model for hurricane evacuation. In: Proceedings of the 87th annual meeting transportation research board, Washington, DC, USA, pp 13–17
Dow K, Cutter SL (2002) Emerging hurricane evacuation issues: hurricane Floyd and South Carolina. Nat Hazards Rev 3(1):12–18
Ferreira CM, Olivera F, Irish JL (2014) Arc StormSurge: integrating hurricane storm surge modeling and GIS. JAWRA J Am Water Resour Assoc 50(1):219–233
Fox SJ, Lachmann M, Meyers L (2021) COVID-19 campus introduction risks for school reopenings. As of Sep 1, 2021. https://interactive.khou.com/pdfs/UT-study-on-COVID-19-school-introduction-risks.pdf
Hijmans R (2019) Geosphere: spherical Trigonometry. R package: version 1:5
Homeland Infrastructure Foundation-Level Data (2020) Hurricane Evacuation Routes API. Last accessed September. https://hifld-geoplatform.opendata.arcgis.com/datasets/hurricane-evacuation-routes/api
Huang SK, Lindell MK, Prater CS, Wu HC, Siebeneck LK (2012) Household evacuation decision making in response to Hurricane Ike. Nat Hazards Rev 13(4):283–296
Hunter E, Mac Namee B, Kelleher JD (2017) A taxonomy for agent-based models in human infectious disease epidemiology. J Artif Soc Soc Simul 20(3)
Jeff Lindner Report on Hurricane Rita (2020). https://cdn.ymaws.com/www.tfma.org/resource/resmgr/Center_Page_News_Brief/Hurricane_Rita.pdf. Accessed Sept 2020
Jelesnianski CP (1992) SLOSH: sea, lake, and overland surges from hurricanes, vol 48. US Department of Commerce, National Oceanic and Atmospheric Administration, National Weather Service

Kang JE, Lindell MK, Prater CS (2007) Hurricane evacuation expectations and actual behavior in Hurricane Lili 1. J Appl Soc Psychol 37(4):887–903

KHOU News (2020) Hurricane Laura: track and spaghetti models. https://www.khou.com/article/weather/hurricane/tropical-spaghetti-models-track-satellite/285-ee04c50b-ce67-4ba4-a18a-075c76956910. Accessed Sept 2021

Knabb RD, Brown DP, Rhome JR (2006) Tropical cyclone report, Hurricane Rita, 18–26 September 2005. National Hurricane Center, 17

Lequime S, Bastide P, Dellicour S, Lemey P, Baele G (2020) Nosoi: a stochastic agent-based transmission chain simulation framework in r. Methods Ecol Evol 11(8):1002–1007

Lindell MK, Lu JC, Prater CS (2005) Household decision making and evacuation in response to Hurricane Lili. Nat Hazards Rev 6(4):171–179

Mesa-Arango R, Hasan S, Ukkusuri SV, Murray-Tuite P (2013) Household-level model for hurricane evacuation destination type choice using hurricane Ivan data. Nat Hazards Rev 14(1):11–20

Modali NK (2005) Modeling destination choice and measuring the transferability of hurricane evacuation patterns. Master's thesis. LSU digital commons. https://digitalcommons.lsu.edu/cgi/viewcontent.cgi?referer=https://scholar.google.com&httpsredir=1&article=4372&context=gradschool_theses

Morrow B, Gladwin H (2005) Hurricane ivan behavioral analysis. Technical Report prepared for the federal emergency management agency and the U.S. Army Corps of Engineers, Wilmington and Mobile Districts

Mwalili S, Kimathi M, Ojiambo V, Gathungu D, Mbogo R (2020) SEIR model for COVID-19 dynamics incorporating the environment and social distancing. BMC Res Notes 13(1):1–5

National Hurricane Center Data in GIS Formats (2020) https://www.nhc.noaa.gov/gis/. Accessed Sept 2020

Palmer M (2001) Using GIS for emergency response. 2001 Esri user conference. Technical paper, number 811

Wilmot CG, Mei B (2004) Comparison of alternative trip generation models for hurricane evacuation. Nat Hazards Rev 5(4):170–178

A Dynamic Social Vulnerability Index to COVID-19 in Mexico

Raúl Sierra-Alcocer, Pablo López-Ramírez, and Graciela González-Farías

Abstract Several months have passed since the appearance of COVID-19, populations that were the most vulnerable at the beginning might not be anymore, and
vice-versa. Government interventions, people behaviours and vaccination policies,
change the social vulnerability. Our work proposes a complementary framework to
the classic vulnerability indexes which aggregate structural variables into composite
indexes. We define a Dynamic Vulnerability Index as an evolving relation between
structural indicators and mortality ratio, we construct this index using a data-driven
approach that updates the mortality ratio and uses Partial Least Squares to find a
weighting of the structural variables at each municipality. Our index is able to distinguish at any given time between zones that are potentially vulnerable but do not
exhibit a high exposure, and zones that are not as vulnerable in terms of their structural
variables but present higher levels of exposure. The southwest part of the country,
comprising the states of Chiapas, Guerrero and Oaxaca, exhibits low Dynamic Vulnerability for most of the study period despite being one of the poorest regions in the
country. This happens because most of the region is relatively isolated and doesn't
have a great influx of people that could carry the virus. On the contrary, the Central
Region where the capital (Mexico City) is located and has been the epicenter of the
pandemic in Mexico, has remained with a high vulnerability for the whole period,
even if it is not particularly poor. Our index represents a complement to the static view
of vulnerability in the context of an evolving pandemic. While static vulnerability
highlights regions that could experience a strong impact, the dynamic vulnerability
highlights regions where there is a strong relationship between the fixed structural
conditions and the evolving epidemic. This complementary picture allows decision
makers to take more focused actions.

R. Sierra-Alcocer (✉)
CONABIO, Ciudad de México, México
e-mail: raul.sierra@conabio.gob.mx

P. López-Ramírez
CentroGeo, Ciudad de México, México
e-mail: pablo.lopez@centrogeo.edu.mx

G. González-Farías
CIMAT, Guanajuato, México
e-mail: farias@cimat.mx

1 Introduction

As of March 2021 there are over 119 million global confirmed COVID-19 cases with 2.6 million associated deaths. More than a year has passed since the initial outbreak in the Hubei province in China, and many countries that successfully coped with a first wave of outbreaks experienced second and even third infection waves requiring new strict containment measures (Hale et al. 2021). This evolution has had different geographical signatures across time: states and provinces that where not affected at the beginning experienced rapid growths in cases and hospitalizations in winter, such is the case of southern Italy or the central region of the United States. As the pandemic evolves, different population groups are exposed to the virus as it spreads from the main urban centers to the less populated rural provinces (Fortaleza et al. 2020); these changes in the exposed population can lead to different mortality outcomes as differences in health care access, general hygiene conditions and socioeconomic status are evidenced, this is specially true in developing countries with already struggling health systems and deep social inequalities.

In Mexico the first recorded case took place in February 2020 and has since passed two million confirmed cases with close to 200,000 deaths (Secretaria de Salud 2021). On March 23, 2020 the Mexican government issued a general stay at home policy that, although not strictly enforced, led to an average mobility reduction of over 50% (Graff-Guerrero et al. 2020). The general stay at home order ended on May 30, 2020. Under these interventions, the epidemiological curve peaked in July 2020 and then entered a slow but steady decline in new cases. The national confinement policy was then substituted by a state level tiered restriction system that allowed individual states to relax or tighten containment measures based on a series of indicators such as the availability of hospital beds and the current trend in confirmed cases. This relaxation and diversification of Non Pharmaceutical Interventions (NPIs) led to a new increase in confirmed cases, hospitalizations and deaths with different impacts across the country. Mexico City saw the worst rise of them during the second wave of the epidemic with peak hospitalizations that surpassed the July peak, while the southwestern states of Chiapas, Oaxaca and Guerrero experienced only mild rises in cases.

Since the beginning of the epidemic in Mexico it has been clear that the profound inequalities in Mexican society translate into differentiated impacts across the country. While some groups are able to follow social distance guidelines and stay at home orders, many other sectors depending on daily wages from informal occupation have not been able to follow policy guidelines. At the same time poverty or precarious living conditions have not been the only discriminant factor in the outcome of the epidemic across Mexico. Other factors such as the economic importance in the national context and the flow of people also have played an important role in the evolution of the epidemic.

The different capabilities of societal groups to cope with hazardous events have been addressed through social vulnerability indexes (Fatemi et al. 2017; United Nations Office for Disaster Risk Reduction 2015). These indexes are based on structural social characteristics: poverty, racial composition, public health access. In a recent article, Acharya and Porwal (2020) propose a social vulnerability index for India, specially tailored to the COVID-19 epidemic, that characterizes vulnerability along five dimensions adapted from the CDC framework (Flanagan et al. 2011): socioeconomic, demographic, housing and hygiene, and availability of individual and community health care. Each dimension is measured through publicly available variables from official sources. Through an aggregation of the percent rankings of each variable for each spatial unit, they summarise the dimensions into an index at the state and department levels, that highlights the different vulnerabilities across the Indian geography. The association between vulnerability, measured through indexes similar to the one proposed by Acharya and Porwal (2020), and disease outcomes is well proven in the literature. Khazanchi et al. (2020) conducted a hierarchical regression that shows that vulnerability (calculated using the same CDC framework) is associated with both, the probability of infection and the severity of the outcomes. Estefania et al. (2020) used a different social vulnerability framework and reach the same set of conclusions. These studies confirm the link between the structural characteristics of societies and provide basis for public policy to address the differentiated impacts of the pandemic across societal groups.

Vulnerability indexes derived from the natural disasters approach are static in nature: exposure to natural disasters is characterized by a single instantaneous occurrence and thus, in this context, vulnerability can be understood as a static concept summarizing the structural characteristics that drive different outcomes from the disaster. However, as articulated in The Lancet (2020), "vulnerability in the present context is a dynamic concept, a person or a group might not be vulnerable at the beginning of the pandemic, but could subsequently become vulnerable depending on the government response". This dynamic nature of vulnerability is not captured by the static indexes, they are ill suited to cope with the complex spatio-temporal dynamics of an evolving pandemic. Government interventions through containment measures or economic incentives to poor populations have a differential impact on vulnerability that is not captured by the static picture. At the same time, static vulnerability indexes, by definition, tend to stress the vulnerability of poor populations, while this is important in planning a strategy to provide help, it also could prove to be prejudicial to these populations if strict containment measures are imposed regardless of whether there is an important risk associated with exposure. In a recent study (Tiwari et al. 2021) developed a novel COVID-19 vulnerability index by combining an impact assessment algorithm with a machine learning approach to include exposure and health outcomes in the vulnerability calculation, while this is an improvement over the classic vulnerability indexes that do not include any measure of exposure or outcomes, it still lacks the temporal dimension.

The dynamic nature of vulnerability becomes more important with the current availability of vaccines. As we are seeing now, vaccine availability varies widely across countries and is specially lacking in developing countries. This limited accessibility is prompting governments to develop immunization strategies to maximize impact through a careful balance of different objectives: minimizing fatalities, maximizing the immunization impact or privileging economic recovery, for example.

From these considerations, we propose a novel Dynamic Social Vulnerability Index (DySoVI) that captures the dynamics of the interaction between the different vulnerability dimensions and the evolution of the epidemic. We depart from a view similar to that of Acharya and Porwal (2020), characterizing vulnerability using the same dimensions proposed by the CDC framework but, instead of aggregating the dimensions through quantile rank aggregation or other static measure such as Principal Component Analysis or additive methods, we use Partial Least Squares (PLS) regression (Höskuldsson 1988) which allows for the explicit incorporation of a target variable in the calculation of composite indexes (Yoon et al. 2015; Yoon and Klasen 2018; Trinchera and Russolillo 2010; Naik and Tsai 2000). To model the Dynamic Vulnerability we use the mortality ratio as the target variable since, on the one hand, it captures both the general exposure of the population to the epidemic and the outcome of this exposure, while on the other hand it is less affected by the testing strategy than the Case Fatality Ratio. This last point is specially important in Mexico where there has been limited testing. We believe that this approach is a valuable addition to both, the current literature on vulnerability to infectious diseases and a useful public policy tool that allows the timely assessment of public policy interventions.

2 Methods

2.1 Dynamic Vulnerability

The concept of vulnerability comes from the disaster management literature. Within this field, risk is the possibility of adverse effects due to a hazardous event (Cardona et al. 2012). Risk derives from the interaction of the actual hazard and the vulnerabilities of the exposed elements. Figure 1 shows a simplified conceptual model for disaster risk management.

Within this framework, vulnerability is understood as the characteristics of an element that make it more prone to suffer the adverse effects of a hazard, while exposure refers to the elements at the sites or areas where the hazard occurs (Birkmann et al. 2006). In general, the four components in Fig. 1 change through space and time, although, for a specific occurrence of a hazardous event such as a hurricane or earthquake, vulnerability can be thought of as static, representing the characteristics of the exposed elements at the time of the event.

Fig. 1 Disaster risk can be understood as the *composition* of three complementary concepts: Hazard, Vulnerability and Exposure. Adapted from Cutter et al. (2003)

This situation is quite different when the hazard is a long lasting phenomenon, like the COVID-19 pandemic. In this case, the picture of vulnerability as a static measure is not sufficient to understand the complex dynamics at play in the spatio-temporal evolution of the hazard. To gain a deeper understanding of the evolution of vulnerability we developed a Dynamic Social Vulnerability Index (DySoVI) as a measure that combines the concepts of risk, exposure and vulnerability into a single index that is able to capture the interplay between the characteristics of the exposed population (vulnerability) with the health outcome (risk).

Our approach to model this interplay is first to select explanatory variables along the same dimensions used in Acharya and Porwal (2020), and then define a target variable that works as a proxy for risk and exposure. In our case, this proxy is the COVID-19 mortality rate. The next step is to aggregate the explanatory variables into a single composite index, for that we chose to use a Partial Least Squares regression (PLS). PLS is similar to Principal Components Analysis, a method that is widely applied in vulnerability research (Aksha et al. 2019; de Loyola Hummell et al. 2016; Frigerio et al. 2018; Uddin et al. 2019). PLS, however, instead of finding the components that maximize the co-variance among the set of explanatory variables, finds the components that maximize the co-variance between the target variable and the set of explanatory variables (Höskuldsson 1988; Sun et al. 2009), in addition, PLS is robust to colinearities in the explanatory variables which is important since we want to include as much information about the structural characteristics of population as possible. A further advantage of PLS is that we can get the relative importance of each explanatory variable through the PLS loadings. This is important in our methodology because changes in the relative importance of the explanatory variables are indicative of changes in the structural characteristics of the exposed population and could guide policy interventions.

By taking the first component of the PLS regression we can construct an index for an arbitrary point in time where mortality ratio is calculated. To adjust a PLS regression for a given day, we use the mortality rate for the previous 28 d of reported data, the PLS loadings gather the relative importance of each explanatory variable to capture social vulnerability to COVID-19 at that point in time.

2.2 Study Design

The Mexican Census provides social and demographic data at the municipal level for the whole country (INEGI 2010), with this data, the National Council for the Evaluation of Social Development Policy (CONEVAL) produces several indicators that measure different dimensions of poverty and lack of services at the municipal level (CONEVAL 2015). To include data about comorbidities known to impact the outcome of COVID-19, we used the small area estimations provided by the National Bureau of Statistics (INEGI 2018). Tables 1, 2, 3 and 4 summarize the data used for our calculations. Data about public and private hospitals was aggregated at the

Table 1 Variables used for the socioeconomic dimension

Socioeconomic dimension

Variable name	Description (source)
Age 6 to 14 not at school	Percentage of population between 6 and 14 years old not attending school (INEGI Population census 2010)
Educational lag	Percentage of population with educational lag (Poverty indicators CONEVAL 2015)
Extreme poverty	Percentage of population living under extreme poverty (Poverty indicators CONEVAL 2015)
Houses with dirt floor	Percentage of houses with dirt floor (INEGI population census 2010)
Illiterate over 15	Percentage of population over 15 years old who are illiterate (INEGI population census 2010)
Lack in healthcare access	Percentage of population lacking adequate health care access (Poverty indicators CONEVAL 2015)
Moderate poverty	Percentage of population living in moderate poverty (Poverty indicators CONEVAL 2015)
Not poor nor vulnerable	Percentage of population that doesn't live under poverty nor is vulnerable (Poverty indicators CONEVAL 2015)
Over 15 with incomplete basic education	Percentage of population over 15 years old with incomplete basic education (INEGI population census 2010)
Place within national context	Position of municipality according to economic importance within the national context (Poverty indicators CONEVAL 2015)
Population under minimum wellness line	Percentage of population living under minimum wellness conditions (Poverty indicators CONEVAL 2015)
Population with at least one lack	Percentage of population lacking in at least one of the four: health access, education, services and social security (Poverty indicators CONEVAL 2015)
Population with at least three lacks	Percentage of population lacking in at least three of the four: health access, education, services and social security (Poverty indicators CONEVAL 2015)
Poverty	Percentage of population living under poverty conditions (Poverty indicators CONEVAL 2015)
Social lag index	Index representing the general social lag for the municipality (Poverty indicators CONEVAL 2015)
Vulnerable due to income	Percentage of population vulnerable to income loss (Poverty indicators CONEVAL 2015)
Vulnerable due to social lag	Percentage of population vulnerable because of social lacks (Poverty indicators CONEVAL 2015)

Table 2 Variables used for the Housing and hygiene dimension

Housing and hygiene dimension	
Households without electricity	Percentage of households without electricity (INEGI population census 2010)
Households without fridge	Percentage of households without electric refrigerator (INEGI population census 2010)
Households without public drainage	Percentage of households without access to public drainage (INEGI population census 2010)
Households without running water	Percentage of households without access to running water (INEGI population census 2010)
Households without washer	Percentage of households without washing machine (INEGI population census 2010)
Lack in house quality	Percentage of households lacking in construction quality (Poverty indicators CONEVAL 2015)
Lack of adequate nutrition	Percentage of households without adequate nutrition (Poverty indicators CONEVAL 2015)
Lack of basic services	Percentage of households without access to basic services: public drainage, running water and electricity (Poverty indicators CONEVAL 2015)

Table 3 Variables used for the Health care dimension

Health care dimension	
Lack in health care access	|Percentage of population lacking access to health care professionals (Poverty indicators CONEVAL 2015)
Lack in social security	Percentage of population without social security services (Poverty indicators CONEVAL 2015)
No health insurance	Percentage of population without health insurance (Poverty indicators CONEVAL 2015)
Pharmacies	Number of pharmacies per population
Private hospital beds	Number of beds in private hospitals per population measured at the state level (CLUES, Health Ministry 2018)
Private medical practice offices	Number of private general medical practice offices (DENUE, INEGI 2018)
Public hospital beds	Number of beds in public hospitals per population measured at the state level (CLUES, Health Ministry 2018)
Public hospitals	Number of public hospitals measured at the state level (CLUES, Health Ministry 2018)
Public medical practice offices	Number of public general medical practice offices (DENUE, INEGI 2018)

Table 4 Variables used for the Epidemiological dimension

Epidemiological dimension	
Prevalence of diabetes	Proportion of people living with diabetes (INEGI small area estimates 2018)
Prevalence of hypertension	Proportion of people living with hypertensive disorders (INEGI small area estimates 2018)
Prevalence of obesity	Proportion of people with obesity (INEGI small area estimates 2018)

Table 5 Selected of intervals for DVI calculation

From February 1st to March 23Th 2020	From the first case to the beginning of the first general confinement measures
From March 23Th to April 30Th 2020	In the middle point of the first general confinement measures
From April 30Th to June 30Th 2020	Around the time of the firs epidemic peak
From June 30Th to August 1st 2020	The descending phase of the first wave
From August 1st to October 15Th 2020	The valley between first and second waves
From October 15Th 2020 to December 1st 2020	The beginning of the second wave
From December 1st 2020 to January 20Th 2021	Around the peak of the second wave
From January 20Th to February 28Th 2021	The last date analyzed

state level since hospitals are concentrated in urban population centers. All variables where recoded so that bigger values imply a higher vulnerability.

As described in Sect. 2.1 our DySoVI measure can be calculated at arbitrary points in time and represents an instantaneous picture of the relationship between the mortality ratio and the variables that characterize vulnerability at each municipality. In order to capture the evolution of this measure we selected seven important moments in the evolution of the epidemic in Mexico (Table 5) and averaged the DySoVI for the intervals defined by those moments.

Data about COVID-19 cases comes from the Ministry of Health of Mexico (SS) (Secretaria de Salud 2021). The SS publishes on a daily basis an updated version of its COVID-19 database. The published data includes records for all the deceased patients who, either have a positive COVID-19 test or are associated with other patient with a positive COVID-19 test; it also includes the municipality of residence of the patient which allows us to compute the mortality ratio. We chose to use mortality instead of case fatality ratio because the testing strategy in Mexico limits the amount of COVID-19 cases detected at any moment (Fernández-Rojas et al. 2021) and mortality statistics are more reliable. In order to account only for the exposure at the time of calculation we don't use the total of accumulated deaths and instead use only patients deceased in the 28 previous days as a proxy for current (high risk) exposure levels.

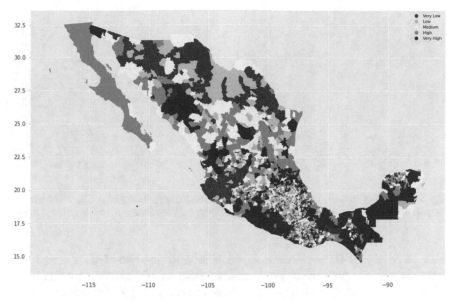

Fig. 2 Social Vulnerability at the municipal level using the methodology from Flanagan (2011)

To construct the DySoVI at the state level we used the average of the municipalities in each state weighted by the proportion of the state population at each municipality. Thus, at the end of the calculations we have a DySoVI for each municipality and state for all intervals listed in Table 5, we also produce the average relative importance of the variables for those same intervals (Fig. 2).

3   Results

Figure 3 shows the evolution of DySoVI and mortality rate for the whole study period together with the selected moments we use to capture the evolution. The first thing to notice is that the DySoVI seems to follow the general temporal pattern of mortality ratio, although the DySoVI shows less inter-state variability. To better understand the difference between mortality rate and DySoVI, in Fig. 4 we show a comparison of the state level rankings for both measures. While Fig. 3 shows that both follow the same general pattern, Fig. 4 shows that DySoVI is effectively capturing a different picture than mortality. For example, Aguascalientes starts with a very low mortality ranking but a high DySoVI ranking, indicating that, due to its structural characteristics, it is at a higher comparative risk. The opposite is true for Morelos, which starts with a high mortality ranking and a relatively low DySoVI, indicating that its internal structure allows it to better cope with the epidemic.

Of special interest are the evolution of Mexico City and Nuevo León compared to those of Chiapas and Guerrero. The former states are the national capital and one of

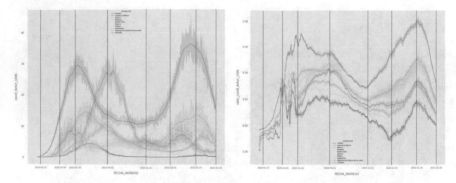

Fig. 3 Comparison between mortality (left) and DySoVI (right) time series. The black vertical lines show the moments selected for analysis. For clarity only ten states are shown in the graphic

the most developed states respectively, while the latter are two of the poorest states in the country. For comparison Fig. 2 presents a map of a Social Vulnerability Index calculated with the method presented in Flanagan et al. (2011). Chiapas and Guerrero are at high vulnerability levels but our DySoVI measure consistently places them at lower ranks since they never experienced considerable surges in mortality ratio and thus exposure and risk are relatively low compared to those of Mexico City and Nuevo León.

To have a comprehensive picture of the spatio-temporal evolution of DySoVI, in Fig. 6 we present a series of maps showing the state level DySoVI for the selected periods. It is interesting to notice that the Southwestern region of Mexico, comprising the states of Chiapas, Oaxaca, Guerrero and Michoacán remains at a lower DySoVI than the North and Central regions. Southwestern states are characterized by higher poverty and social lacks than the Central and Northern regions, but their relative isolation and the types of economic activities seem to have spared them from the worst consequences of the epidemic.

Another important feature of our DySoVI measure is its ability to track the evolution of the relative importance of the different structural characteristics involved in its calculation. In Fig. 5 we show the evolution of the rankings for the variables we used along the four vulnerability dimensions. Within the socioeconomic dimension it is relevant to notice that both, the population with at least three social lacks and the position within the national context increase their relative importance with time, while most variables relating with poverty and health access decrease their importance. This does not necessarily mean that poverty and health care access are not important but rather that they don't help in discriminating places severely affected by the epidemic (Fig. 6).

Within the Housing and Hygiene dimension the lack of basic services, house quality, and lack of running water are consistently ranked high, while public drainage and access to electricity are the lowest. Again, this is a consequence of the characteristics of the populations exposed to the epidemic. The Health care dimension shows that, while at the beginning of the epidemic the most important variable was lack of

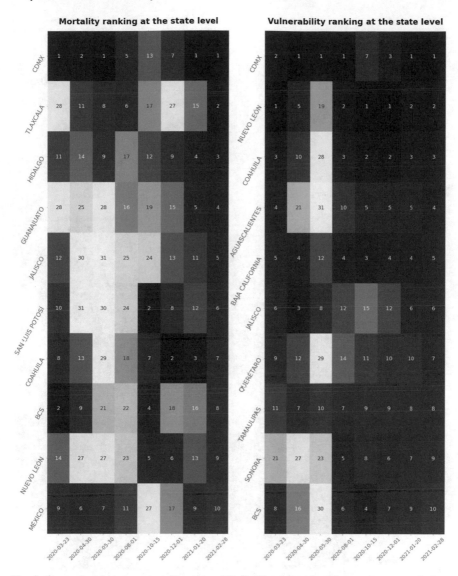

Fig. 4 Comparison between mortality rate and DySoVI rankings for the top ten states at the last analysis date for the selected time periods. Numbers inside the colored squares show the ranking of the state at each interval

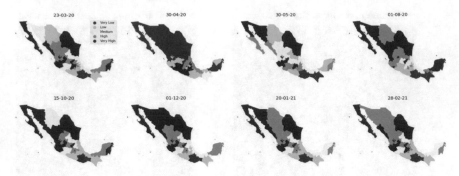

Fig. 5 Evolution of state-level DVI

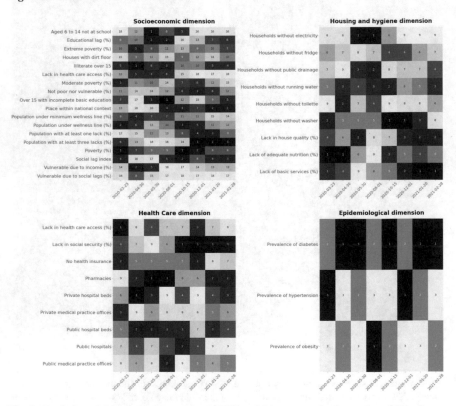

Fig. 6 Evolution of the relative importance of DVI variables across the four vulnerability dimensions

health care access, as the epidemic evolved lack of social security became the most important variable together with the amount of pharmacies. Lack of social security is strongly associated with informality in Mexico (Brandt 2011) which might explain its relevance since informality is in turn associated with the incapacity of populations to follow stay at home advice (Salamanca and Vargas 2020). The importance of pharmacies might be explained by the dominance of urban populations affected by the epidemic.

The Epidemiological dimension shows that the prevalence of diabetes is the most important factor for most of the time, while obesity and hypertension switch places. The importance of diabetes is consistent with what is known about the association of comorbidities with higher mortality risk from COVID-19 (Hernández-Garduño 2020; Parra-Bracamonte et al. 2020) and the high prevalence among Mexican population (Meza et al. 2015).

Finally, our DySoVI is calculated at the municipal level, which allows us to inspect the local inhomogeneities on the DySoVI at the state level. For example, in Fig. 7 we show a series of maps for the evolution of DySoVI at the municipal level for the states of Chiapas, a state showing low DySoVI for the whole period but with high poverty, and Nuevo León (Fig. 7b), which is one of the most developed states and exhibits a high DySoVI. These maps shows us that in both cases the DySoVI along the municipalities is far from homogeneous. In Chiapas the coastal municipalities and the central area where the state capital is located exhibit the greatest DySoVI for the whole period, while the central mountainous region, predominantly poor and deprived, exhibits a low DySoVI. In Nuevo León, the central region, where the state capital is located remains at a high DySoVI level for the whole period, but there are pockets of high DySoVI at the southern part of the state and in the north, at the border with the United States, at specific points in time.

4 Discussion

The response efforts to tackle the COVID-19 epidemic in Mexico need accurate and timely data to update resource allocation and containment measures. The static vulnerability picture presented by indexes such as the one proposed by Acharya and Porwal (2020) allow for an initial resource allocation and planning, but the complex interaction between the spatio-temporal spread of the epidemic and the structural characteristics of population calls for a more nuanced vulnerability approach that is able to capture this interplay. Our DySoVI measure, by capturing the evolution of the relationship between the exposed population, the health outcomes and the structural characteristics of the population, is able to provide this dynamic picture, allowing the identification of vulnerable places that would be otherwise missed. For example, most of Northern Mexico would be put under low static vulnerability (Fig. 2), but from our analysis it is clear that it is one of the most vulnerable regions most of the time. The contrary is also true, while most of the southwestern region exhibits a high static vulnerability, our DySoVI consistently places it at the lowest levels. This

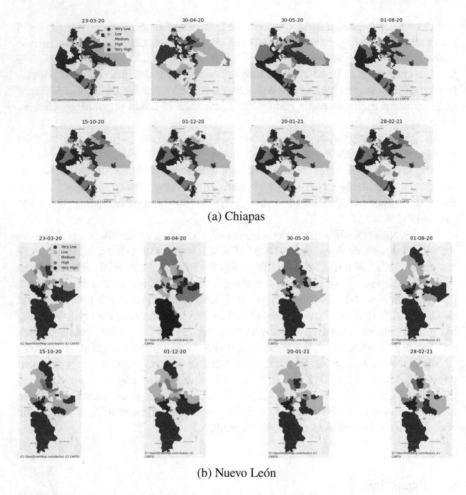

(a) Chiapas

(b) Nuevo León

Fig. 7 DVI at the municipal level for the states of Chiapas and Nuevo León

fine grain spatial and temporal resolution allows policy makers to make informed decisions about containment measures and resource allocation at different points in time and space, thus optimizing both, the overall stringency of NPIs which may have a high economic impact, and the efficient management of health resources.

Another advantage of our approach is that PLS combines the vulnerability variables in such a way as to maximize the relationship between them and the mortality ratio. Furthermore, our conceptual approach combining exposure, risk and vulnerability into a single index, circumvents the conflation between vulnerability and susceptibility discussed in Acharya and Porwal (2020) by explicitly combining both concepts, thus, our DySoVI measures both, susceptibility and vulnerability at the same time.

While our DySoVI has the aforementioned advantages, they also come at a cost. The most evident drawback is in interpretability. Indexes based on the CDC framework have the advantage of being easily interpretable and replicable, while our PLS based DySoVI is certainly more complex and involved and might not be as easily interpretable to decision makers. Another limitation is that it assumes that the same processes drive vulnerability for urban and rural populations which might not be the case. Rural environment in Mexico, specially in the poorest regions, is characterized by small communities relatively disconnected from larger population centers. Mortality rates and comorbidities prevalence at these places are more likely to be underrepresented because of the lack of health care facilities.

Our DySoVI measure is a useful tool in the development of local and time dependant interventions and resource allocation, it allows policy makers to make fine grained spatial and temporal decisions that distinguish between places with high static vulnerability but not seriously affected by the epidemic and places that are at relative low static vulnerability levels but with a high exposure. This improvement allows for a better resource allocation and mitigation of the economic burden of NPIs.

In conclusion, following a data driven methodology, we developed a novel dynamic vulnerability index to COVID-19 for Mexico. The approach can be reproduced and extended for other countries and re-adjusted at any time with new evidence. Our main contribution is hence twofold, on the one hand our work represents an improvement over traditional static vulnerability indexes since it is able to capture the evolution of the relationship between the structural characteristics of the population and the exposure and associated risk, and on the other hand, it goes beyond indexes adjusted from a combination of static explanatory data and expert knowledge, actually relating the epidemic outcomes to the explanatory variables. The evolution of the pandemic shows that vulnerability must be understood as a dynamic concept. As the pandemic evolves governments and people react and effectively modify the vulnerability landscape. Finally, we live on an era of data, and this pandemic is no exception (Ienca and Vayena 2020), the amount of data that humanity is gathering around this pandemic is unprecedented, and we need methods that help us make sense of the vast amount of data. As new data is generated everyday, and with it, new information, we should be prepared to use this new information as quickly as possible.

References

Acharya R, Porwal A (2020) A vulnerability index for the management of and response to the COVID-19 epidemic in India: an ecological study. Lancet Global Health 0(0):1–10. ISSN 2214109X. 10.1016/S2214-109X(20)30300-4. https://linkinghub.elsevier.com/retrieve/pii/S2214109X20303004

Aksha SK, Juran L, Resler LM, Zhang Y (2019) An analysis of social vulnerability to natural hazards in Nepal using a modified social vulnerability index. Int J Disas Risk Sci 10(1):103–116. ISSN 2095-0055, 2192-6395. https://doi.org/10.1007/s13753-018-0192-7

Birkmann J, Dech S, Hirzinger G, Klein R, Klüpfel H, Lehmann F, Mott C, Nagel K, Schlurmann T, Setiadi NJ, Siegert F, Strunz G (2006) Measuring vulnerability to promote disaster resilient societies? Conceptual frameworks and definitions. In: Measuring vulnerability to natural hazards: towards disaster resilient societies. UNU-Press, Tokio

Brandt N (2011) Informality in Mexico. Working Paper 896, OECD, Paris, October

Cardona O-D, van Aalst MK, Birkmann J, Fordham M, McGregor G, Rosa P, Pulwarty RS, Schipper ELF, Sinh BT, Décamps H, Keim M, Davis I, Ebi KL, Lavell A, Mechler R, Murray V, Pelling M, Pohl Smith A-O, Thomalla F (2012) Determinants of risk: exposure and vulnerability. In: Field CB, Barros V, Stocker TF, Dahe Q (eds) Managing the risks of extreme events and disasters to advance climate change adaptation. Cambridge University Press, Cambridge, pp 65–108. 978-1-139-17724-5. https://doi.org/10.1017/CBO9781139177245.005

CONEVAL (2015) Pobreza municipal 2010–2015. https://www.coneval.org.mx/Medicion/Paginas/Pobreza-municipal.aspx. Accessed 01 Oct 2020

Cutter SL, Boruff BJ, Shirley WL (2003) Social vulnerability to environmental hazards [*]: social vulnerability to environmental hazards. Soc Sci Q 84(2):242–261. ISSN 00384941. 10.1111/1540-6237.8402002

de Loyola Hummell BM, Cutter SL, Emrich CT (2016) Social vulnerability to natural hazards in Brazil. Int J Disaster Risk Sci 7(2):111–122. ISSN 2095-0055, 2192-6395. https://doi.org/10.1007/s13753-016-0090-9

Farin Fatemi, Ali Ardalan, Benigno Aguirre, Nabiollah Mansouri, and Iraj Mohammadfam. Social vulnerability indicators in disasters: Findings from a systematic review, 6 2017. ISSN 22124209

Fernández-Rojas MA, Esparza MAL-R, Campos-Romero A, Calva-Espinosa DY, Moreno-Camacho JL, Langle-Martínez AP, García-Gil A, Solís-González CJ, Canizalez-Román A, León-Sicairos N, Alcántar-Fernández J (2021) Epidemiology of COVID-19 in Mexico: symptomatic profiles and presymptomatic people. Int J Infect Dis 104:572–579. ISSN 1201-9712. 10.1016/j.ijid.2020.12.086

Flanagan BE, Gregory EW, Hallisey EJ, Heitgerd JL, Lewis B (2011) A social vulnerability index for disaster management a social vulnerability index for disaster management. J Homel Secur Emerg Manag. https://doi.org/10.2202/1547-7355.1792

Fortaleza CMCB, Guimarães RB, De Almeida GB, Pronunciate M, Ferreira CP (2020) Taking the inner route: spatial and demographic factors affecting vulnerability to COVID-19 among 604 cities from inner São Paulo State, Brazil. Epidemiol Infect 148. ISSN 14694409. 10.1017/S095026882000134X. https://doi.org/10.1017/S095026882000134X

Frigerio I, Carnelli F, Cabinio M, De Amicis M (2018) Spatiotemporal pattern of social vulnerability in Italy. Int J Disaster Risk Sci 9(2):249–262. ISSN 2095-0055. 2192-6395. https://doi.org/10.1007/s13753-018-0168-7

Hale T, Angrist N, Goldszmidt R, Kira B, Petherick A, Phillips T, Webster S, Cameron-Blake E, Hallas L, Majumdar S, Tatlow H (2021) A global panel database of pandemic policies (Oxford COVID-19 Government Response Tracker). Nat Hum Behav 1–10. ISSN 2397-3374. 10.1038/s41562-021-01079-8

Hernández-Garduño E (2020) Obesity is the comorbidity more strongly associated for Covid-19 in Mexico. A case-control study. Obes Res Clin Pract 14(4):375–379. ISSN 1871-403X. 10.1016/j.orcp.2020.06.001

Höskuldsson A (1988) PLS regression methods. J Chemom 2(3), 211–228. ISSN 0886-9383, 1099-128X. https://doi.org/10.1002/cem.1180020306

Ienca M, Vayena E (2020) On the responsible use of digital data to tackle the COVID-19 pandemic. Nat Med 26(4):463–464. ISSN 1546-170X. 10.1038/s41591-020-0832-5

INEGI (2010) Censo nacional de población y vivienda. https://inegi.org.mx/programas/ccpv/2010/. Accessed 01 Oct 2020

INEGI (2018) Prevalencia de obesidad, hipertensión y diabetes para los municipios de méxico. https://www.inegi.org.mx/investigacion/pohd/2018/. Accessed 01 Feb 2021

Khazanchi R, Beiter ER, Gondi S, Beckman AL, Bilinski A, Ganguli I (2020) County-level association of social vulnerability with COVID-19 cases and deaths in the USA, vol 6. ISSN 15251497. https://link.springer.com/article/10.1007/s11606-020-05882-3

Lara-Garcia OE, Retamales VA, Suarez OM, Parajuli P, Hingle S, Robinson R (2020) Application of social vulnerability index to identify high- risk population of contracting COVID-19 infection: a state-level study. https://doi.org/10.1101/2020.08.03.20166983

Mario Graff-Guerrero, Sánchez-Siordia Oscar, Daniela Moctezuma, Eric Tellez, Miranda Sabino (2020) Medición de movilidad usando facebook, google y twitter. Technical report, CONACyT

Meza R, Barrientos-Gutierrez T, Rojas-Martinez R, Reynoso-Noverón N, Palacio-Mejia LS, Lazcano-Ponce E, Hernández-Ávila M (2015) Burden of type 2 diabetes in Mexico: past, current and future prevalence and incidence rates. Prev Med 81:445–450. ISSN 0091-7435. 10.1016/j.ypmed.2015.10.015

Naik P, Tsai C-L (2000) Partial least squares estimator for single-index models. J R Stat Soc: Ser B (Stat Methodol) 62(4):763–771. ISSN 1369-7412, 1467-9868. https://doi.org/10.1111/1467-9868.00262

Parra-Bracamonte GM, Lopez-Villalobos N, Parra-Bracamonte FE (2020) Clinical characteristics and risk factors for mortality of patients with COVID-19 in a large data set from Mexico. Annals Epidemiol 52:93–98.e2. ISSN 1047-2797. 10.1016/j.annepidem.2020.08.005

Salamanca JDG, Vargas G (2020) Quarantine and informality: reflections on the colombian case. Space Cult 23(3):307–314. ISSN 1206-3312, 1552-8308. https://doi.org/10.1177/1206331220938626

Secretaria de Salud. Covid-19, 2021. data retrieved from Secretaria de Salud. https://www.gob.mx/salud/documentos/datos-abiertos-152127

Sun L, Ji S, Yu S, Ye J (2009) On the equivalence between canonical correlation analysis and orthonormalized partial least squares. In: Proceedings of the 21st international jont conference on artifical intelligence, IJCAI'09, pp 1230–1235, San Francisco, CA, USA, July 2009. Morgan Kaufmann Publishers Inc

The Lancet. Redefining vulnerability in the era of COVID-19. Lancet 395(10230):1089. ISSN 01406736. 10.1016/S0140-6736(20)30757-1. https://www.uneca.org/sites/. https://linkinghub.elsevier.com/retrieve/pii/S0140673620307571

Tiwari A, Dadhania AV, Ragunathrao VA, Oliveira ER (2021) Using machine learning to develop a novel COVID-19 vulnerability index (C19VI). Sci Total Environ 773:145650. ISSN 0048-9697. 10.1016/j.scitotenv.2021.145650

Trinchera L, Russolillo G (2010) On the use of structural equation models and pls path modeling to build composite indicators. University of Macerata, Italy

Uddin MN, Islam AS, Bala SK, Islam GT, Adhikary S, Saha D, Haque S, Fahad MG, Akter R (2019) Mapping of climate vulnerability of the coastal region of Bangladesh using principal component analysis. Appl Geogr 102:47–57. ISSN 01436228. 10.1016/j.apgeog.2018.12.011

United Nations Office for Disaster Risk Reduction (2015) Sendai framework for disaster risk reduction 2015–2030. In: UN world conference on disaster risk reduction, p 37, Sendai, Japan, 2015. United Nations Office for Disaster Risk Reduction

Yoon J, Klasen S (2018) An application of partial least squares to the construction of the Social Institutions and Gender Index (SIGI) and the Corruption Perception Index (CPI). Soc Indic Res 138(1):61–88. ISSN 0303-8300, 1573-0921. https://doi.org/10.1007/s11205-017-1655-8

Yoon J, Klasen S, Dreher A, Krivobokova T (2015) Composite indices based on partial least squares. Discussion Papers 171, Georg-August-Universität Göttingen, Courant Research Centre - Poverty, Equity and Growth (CRC-PEG), Göttingen

Effects of COVID-19 in Mexico City: Street Robbery and Vehicle Theft Spatio-Temporal Patterns

Ana J. Alegre-Mondragón and Cristian Silva-Arias

Abstract As a result of the changes in social behavior due to lockdown measures aimed to avoiding COVID-19 infection, changes in crime patterns have been observed in several cities around the world. This study has two objectives: (1) Analyze the spatio-temporal patterns of the incidence of street robbery and vehicle theft in Mexico City, before and after the social distancing measures begun. Throughout this period, it has been shown a decrease in high-impact robberies in Mexico City. However, changes in spatial patterns have not been studied yet. (2) Propose an algorithm for the visualization of spatio-temporal relationships of crimes to identify near repeat patterns. These two objectives are considered relevant to identify areas of repeat victimization, especially before an imminent return to routine activities in the city, such as the return to school, the reopening of restaurants, movie theaters, shopping malls and other businesses; and thus be able to contribute to identify and prevent these crimes. One of the main results is that despite crime volumes decreased, some specific crime locations remained after the lockdown.

1 Introduction

In an attempt to reduce the number of person-to-person infections and to mitigate the spread of the virus, on March 24, 2019, the Mexican federal government published the agreement establishing the preventive measures in order to mitigate and control the health risks posed by the SARS-CoV2 virus disease (COVID-19). Some of the measures decreed included: avoiding attendance to work centers and public spaces, suspending school activities, canceling mass events, closing non-essential businesses. In this context, the change in routines experienced by many cities in the world because of the pandemic can be seen as an opportunity for social experiments to conduct

A. J. Alegre-Mondragón (✉) · C. Silva-Arias
Centro de Investigación en Ciencias de Información Geoespacial, Contoy 137, Col. Lomas de Padierna, Alcaldía Tlalpan, 14240 México, CDMX, México
e-mail: jalegre@centrogeo.edu.mx

C. Silva-Arias
e-mail: csilva@centrogeo.edu.mx

© The Author(s), under exclusive license to Springer Nature Switzerland AG 2022
R. Tapia-McClung et al. (eds.), *Advances in Geospatial Data Science*, Lecture Notes in Geoinformation and Cartography, https://doi.org/10.1007/978-3-030-98096-2_14

studies, including crime behavior patterns, "...these orders impacted countries, states, and communities at different times and in different ways, a naturally occurring, quasi-randomized control experiment has unfolded, allowing the testing of criminological theories as never before." (Ben and Marcus 2020).

The study presented in this document is relevant due to the scenario faced in different cities around the world due to the difficulties as a consequence of the COVID-19 pandemic. At the time of this document's publication, Mexico has experienced a health crisis with more than 258 thousand deceases in the country of which 37 thousand were registered in Mexico City up to August 29, 2021, a drop in GDP of 8.5% in 2020 in the country according to the results of the National Institute of Statistics and Geography (INEGI) (Instituto Nacional de Estadística y Geografía (INEGI) 2021), an increase in unemployment that has caused half a million Mexicans to lose their jobs at the same time that informal commerce is on the rise by 7.7% (El País 2015).

It is important to highlight that numerous studies have already been conducted on the effects of the COVID-19 pandemic on crime in different cities around the world. Most of these studies have analyzed trends in different types of crime and their behavior over time, for example, for the United States: Ashby Matthew (2020), Ashby (2020), Mohler et al. (2020), Campedelli et al. (2020), Abrams (2021), Alex R Piquero et al. (2020); in Vancouver, Canada Hodgkinson and Andresen (2020); in Sweden, Gerell et al. (2020). The results of these studies coincide in that in some cities there is a decrease in robberies and common crimes, while crimes against life such as homicides did not show the same trend. Some of these studies show domestic violence increased, for example in Los Angeles and Dallas.

There are recent studies on changes in spatial patterns of crime due to the pandemic, in Chicago Campedelli et al. (2020) and Yang et al. (2021) analyzed changes in spatio-temporal patterns over time, the main findings in this city is that significant changes during 2020 were detected in crime distributions (Yang et al. 2021) and furthermore, crime trends do not behave in the same way across different areas of a city and across different types of crime (Campedelli et al. 2020). It is noted a regional study in Australia which concludes that "regional differences most likely resulted from differences in local demographic, economic and criminal opportunity structures" (Payne et al. 2021). In London, Sun et al. (2021) have among their main findings that "violence-against-the-person" rate has no statistically significant association with COVID-19 infection rate; and robbery rate, burglary rate, and theft and handling rate have a statistically significant and negative association with COVID-19 infection rate in both April and May", while in Makassar City, Indonesia, Syamsuddin Rahman (2021) detected an increase in the number of theft offenses of 42.5%, contrary to the results seen in other cities, but what is most relevant is that the places of concentration of these crimes changed, they decreased in the central and financial areas of the city, while they increased in residential areas, these works coincide in that they analyze areas, none of them analyze crime patterns at the point level.

In Mexico City, Estevez-Soto (2021) and Balmori et al. (2021) analyzed the change due to social distancing measures in the volume trends and time series behavior of the different types of crime committed in the city, in both studies they concluded that there was a significant decrease in common crimes, but Balmori (2021) emphasized that crimes that may be more related to organized crime did not decrease, the pandemic had no effect on trends in homicides, kidnappings and extortion. However, there are no studies yet that analyze the spatio-temporal patterns changes of crime for Mexico City because of lockdown, it is important to conduct studies examining possible changes in spatial-temporal patterns before and after pandemic social distancing measures and changes in routine activities that may lead to public policy recommendations aimed at crime prevention at certain times and in specific areas of the city. One of the main findings is that, while the amount of crime decreased, some specific locations remain concentrated crime hotspots in the city.

This work aims to analyze the concentration patterns of street robbery and vehicle theft in Mexico City before and after the isolation and social distancing measures due to the COVID-19 pandemic. In addition, an algorithm is proposed to identify and visualize spatio-temporal patterns of crime at a specific level for Mexico City, this algorithm can be replicated in other cities and it is aimed to make a difference by finding repeated victimization hot spots beyond administrative boundaries such as police responsibility areas.

The document is organized as follows: the second section reviews the theories that support repeated victimization. The third section details the data, then explains the methodology used to model the algorithm that identifies priority spatio-temporal patterns in Mexico City. In the fifth section, the results for the crimes of vehicle theft and street robbery are discussed. Finally, conclusions about public policy recommendations for crime prevention, limitations and suggestions for future studies are presented.

2 Background: The Routine Activities Theory

The routine activities theory explains the occurrence of criminal acts as a confluence of several circumstances. First, there must be a motivated offender. Second, there must be a desirable target. Third, the target and the offender must be in the same place at the same time. Finally, three other types of controllers (intimate controllers, guardians, and place managers) must be absent or ineffective (Eck and Weisburd 2015).

According to Cohen and Felson (1979) most criminal acts require the convergence in time and space of likely offenders, suitable targets, and the absence of guardians, and the social structure produces this convergence during everyday life. Now, due to the measures taken by the different governments to prevent COVID-19 contagions, these routines were broken and therefore caused the crime patterns to change as well. The fact that many people stayed at home meant that "desirable targets" were less available or that motivated offenders had to move elsewhere to find these targets. At the same time, it may be caused because the priorities of the authorities changed during this period and there were also fewer police or guards on duty to prevent crime from occurring.

3   Data

This section presents the data that were used, the available sources of information and a statistical description of their behavior. Its first subsection relates to the data on the behavior of mobility in Mexico City and the selection of the study period, and its second subsection concerns the data for high-impact robberies.

3.1   Mexico City Mobility and the Period Under Study

In this document, two sources of information were reviewed to observe the changes in mobility in Mexico City during the lockdown for the COVID-19 pandemic. The first of these sources was from Apple[1] and the second one was from Google.[2] In both cases a first decrease is identified from the middle of March, despite the fact that the Federal Government Agreement was published on March 24 (Secretaría de Salud del Gobierno de los Estados Unidos Mexicanos 2019) Mexico City changed its mobility from March 16th. Therefore, in this paper we take this date as the point of change of routine activities in the city and then from this point data was taken for one year before and one year after for the analysis.

3.2   High-Impact Robbery

The data source reviewed to obtain information used to analyze the behavior of high-impact robberies in Mexico City was the investigation files published by the City Government available through the City Government's open data portal,[3] this type of data has limitations since they are only the complaints that are generated in the public ministries and implies a very high black figure,[4] however these records are classified using a catalog that allow for easy distinction of the high impact robberies reported.

After a first analysis of the data, it can be seen that violent and non-violent street robberies and violent and non-violent vehicle theft are the crimes with the highest number of investigation files, both representing 70% of all high-impact robberies during the study period. Vehicle theft in Mexico City is one of the crimes with a lower than average crime rate, in 2020 it was 39%, this means it is reported more

---

[1] https://covid19.apple.com/mobility.

[2] https://www.google.com/covid19/mobility/.

[3] https://datos.cdmx.gob.mx/.

[4] According to the results of the National Survey of Victimization and Perception on Public Safety (ENVIPE-2020) the overall black figure for Mexico City is 94%, this means that it is the proportion of crimes committed in which there was no complaint or an investigation folder was not initiated during 2019.

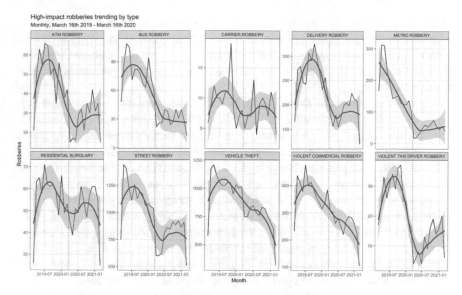

Fig. 1 High impact robbery trending by type. *Source* Elaborated by the authors based on information from the investigation files https://datos.cdmx.gob.mx/

frequently than the rest of the crimes. According to Vilalta vehicle theft is one of the most economically damaging common crimes, since it generates a cost not only for the owners and family expenses, but also for the insurance companies (Vilalta Perdomo 2011). Additionally, street robbery is one of the most difficult crimes to study, there is little literature developed on this subject and it is a crime that is difficult to treat and prevent, however, it is one of the crimes that generates the most complaints to the authorities in Mexico City. The following graphs (Fig. 1) show the changes in trends for all high-impact robberies before and after March 16th, 2020.

4 Methods

4.1 Near Repeat Analysis

This type of analysis is used to detect patterns of interactions in space and time of specific types of crimes, with the aim of finding not only places where crimes are committed more frequently, but also in certain periods of time, identifying these patterns has become useful for prioritizing areas where it is necessary to concentrate resources or develop prevention measures (Philip and Michael 2017; Zengli and Xuejun 2017).

According to Ratcliffe the process of accurately distinguishing the repeat locations has always been difficult. One of the main findings is that the period of greatest risk is immediately after the first theft (Ratcliffe and McCullagh 1998). However, with the development of Geographic Information Systems, now it is easier to develop algorithms to recognize statistically significant time periods and distances for different types of crime. Identifying these patterns is very useful for preventing crime in specific locations.

Many researchers have found that a location where one crime occurred is very likely to be the target of another crime within close periods of time and distance. Johnson et al. (1997) used this technique to find a pattern of repeat home burglaries within a range of 1–2 blocks and 1–2 weeks after the first burglary in a residential area of Merseyside. The patterns change depending on the type of crime being analyzed due to the different mechanics involved in their occurrence, however the principle of the potential re-victimization of a previously occurring crime is preserved.

Near repeat analysis is used as a proactive policing practice to prevent future crimes and to make better use of available resources by focusing police actions in specific areas. A number of tools and resources are currently available to estimate spatial and temporal patterns to identify near-repeat patterns. The analysis made in this document used the Near Repeat package for R (Steenbeek 2018), uses geographic data to quantify the spatio-temporal associations between events using the Knox test and Monte Carlo simulations to estimate the probabilities that an event will recur after a previous one.

4.2 Building of Spatio-Temporal Relationship Lines

In order to facilitate the identification of areas where there are patterns of repeat offenses, the R programming language was used to process the geographic data and visualize the relationships between places where crimes occurred. In this case, two experiments were carried out using the available data on street robbery and vehicle theft to compare the results obtained one year before and one year after lockdown. From the patterns identified through near repeat analysis, the number of days and distance in which the probability of repeat offenses is high is identified. These parameters were used to establish the thresholds at which the space and time relationships described below were built.

The methodology proposed to visually represent repeat victimization consists of selecting a geographic point where a crime was reported, searching for the surrounding points closest in time and space within the established thresholds by near repeat analysis and connecting them with a line which is called the "spatio-temporal relationship line". This procedure is repeated for each of the points within the dataset and those points that had already been previously connected are omitted to avoid duplication, obtaining a multiline that covers an area where there is a pattern of repetition.

Then, each spatio-temporal relationship line is assigned a weight by counting the number of connected points, which is classified into 4 categories using Fisher's algo-

Fig. 2 Nearest points by distance(left) and Nearest points by time (right). *Source* Elaborated by the authors

rithm,[5] where categories are assigned from lowest to highest weights. The multilines are visualized on a map omitting those with the least interaction (labeled as category 1), in which category 4 is the one with the highest number of spatio-temporal interactions, and allowing the identification of the places where there is an important repeated victimization pattern of the selected crime (Fig. 2).

5 Results

The time series plot confirm that there has been a decrease after the confinement due to the pandemic. In the following two sections the findings are shown, the first one for street robbery and the second one for vehicle theft.

5.1 Street Robbery

According to the near repeat analysis in the case of street robbery, significant patterns indicate that there is a high probability (between 20% and 57%) of repeated victimization at a distance of up to 450 m, during the first three days of a previous robbery (Ratcliffe et al. 2008). The area of highest concentration is Downtown which is a highly commercial and busy area, after March 16, 2020 similar patterns were found in the same locations, but with less intensity. A relevant finding of this experience is that these patterns had already been identified since 2009, during an academic collaboration between the Mexico City Police and the CentroGeo (Martínez-Viveros et al. 2013). In other words, the patterns of concentration in time and space have been repeated in these places: Downtown, Tacuba and Tacubaya.

---

[5] Fisher's algorithm allows to calculate exactly an optimal partition from a continuous numerical variable into a given number of classes. Unlike other classification methods such as standard deviations, quartiles or Jenks' natural breaks, this method allows to obtain a better categorization for this case.

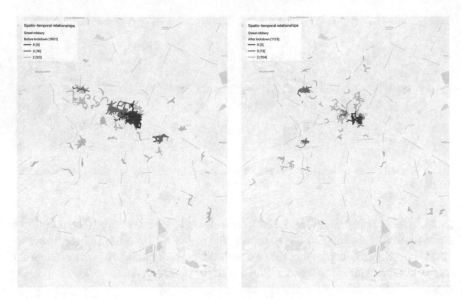

Fig. 3 Street Robbery: Spatio-temporal relationship lines before march 16th (left) and Spatio-temporal relationship lines after march 16th (right). *Source* Elaborated by the authors

5.2 Vehicle Theft

According to the results obtained from the near repeat analysis for vehicle theft, relevant patterns were identified and show a high probability (between 20% to 50%) of further thefts occurring within a distance of up to 150 m in a time range between 6 and 12 d after a first theft was committed. Vehicle theft shows more dispersed spatio-temporal interaction patterns across the city. Some of the most relevant areas were found near the entrances and the exits of the city and on main avenues (Figs. 3 and 4).

6 Conclusions

This study is a first approach to the spatio-temporal behavior of two of the main crimes reported in Mexico City, vehicle theft and street robbery, however for future work it would be useful to incorporate another types of environmental variables that may be related to this type of crime, for example the location of subway stations, transit station terminals with connections to the Mexico State (Estado de México) for street robbery, on the other hand the location of parking lots, residential centers with high population density and lack of closed parking lots and shopping malls for vehicle theft. The results are consistent with other studies, after the pandemic lockdown the robberies in general decreased in volume, but some of the spatio-temporal

Fig. 4 Vehicle Theft: Spatio-temporal relationship lines before march 16th (left) & Spatio-temporal relationship lines after march 16th (right). *Source* Elaborated by the authors

patterns remained in the same places, therefore, the locations indicated may be relevant to prevent future crimes and address policing strategies focusing on specific places and optimizing resources distribution.

Supplementary Information

The algorithm that was developed for the visualization of the spatio-temporal interactions resulting in this study is available in the GitHub repository located at https://github.com/cris-silva/nra-cdmx-covid19.

Acknowledgements We would like to thank Camilo Caudillo Cos, Rodrigo Tapia McClung, Elvia Martínez Viveros and José Ignacio Chapela Castañares (CentroGeo), whose previous work for Mexico City made it possible to follow up and prepare this article. Finally we like to thank José Luis Silván Cárdenas and Carlos Javier Vilalta Perdomo.

References

Abrams DS (2021) Covid and crime: an early empirical look. J Public Econ 194:104344
Ashby Matthew PJ (2020) Initial evidence on the relationship between the coronavirus pandemic and crime in the united states. Crime Sci 9:1–16
Ashby MPJ (2020) Changes in police calls for service during the early months of the 2020 coronavirus pandemic. Polic: A J Policy Pract 14(4):1054–1072

Campedelli GM, Aziani A, Favarin S (2020) Exploring the immediate effects of covid-19 containment policies on crime: an empirical analysis of the short-term aftermath in Los Angeles. Am J Criminal Justice 1–24

Campedelli GM, Favarin S, Aziani A, Piquero AR (2020) Disentangling community-level changes in crime trends during the covid-19 pandemic in Chicago. Crime Sci 9(1):1–18

Cohen LE, Felson M (1979) Social change and crime rate trends: a routine activity approach. Am Sociol Rev 588–608

De la Miyar JR, Hoehn-Velasco L, Silverio-Murillo A (2021) Druglords don't stay at home: Covid-19 pandemic and crime patterns in Mexico city. J Criminal Justice 72:101745

Eck J, Weisburd DL (2015) Crime places in crime theory. Crime and place: Crime prevention studies, vol 4

El País, periódico El desempleo en México crece en más de medio millón de personas en un año. https://elpais.com/mexico/2021-05-27/el-desempleo-en-mexico-crece-en-mas-de-medio-millon-de-personas-en-un-ano.html, Micaela Varela. 20 de mayo, 2015

Estévez-Soto PR (2021) Crime and covid-19: effect of changes in routine activities in Mexico city. Crime Sci 10(1):1–17

Gerell Manne, Kardell Johan, Kindgren Johanna (2020) Minor covid-19 association with crime in Sweden. Crime Sci 9(1):1–9

Glasner Philip, Leitner Michael (2017) Evaluating the impact the weekday has on near-repeat victimization: a spatio-temporal analysis of street robberies in the city of vienna, austria. ISPRS Int J Geo-Inf 6(1):3

Hodgkinson T, Andresen MA (2020) Show me a man or a woman alone and i'll show you a saint: changes in the frequency of criminal incidents during the covid-19 pandemic. J Criminal Justice 69:101706

Instituto Nacional de Estadística y Geografía (INEGI) (2021) Producto Interno Bruto de México Durante el Cuarto Trimestre de 2020 *Comunicado de Prensa Núm. 157/21*, 25 de febrero 2021

Johnson SD, Bowers K, Hirschfield A (1997) New insights into the spatial and temporal distribution of repeat victimization. B J Criminol 37(2):224–241

Martínez-Viveros E, Chapela JI, Morales-Gamas A, Caudillo-Cos C, Tapia-McClung R, Ledesma M, Serrano F (2013) Construction of a web-based crime geointelligence platform for Mexico city's public safety, Springer. Crime Modeling and Mapping Using Geospatial Technologies, pp 415–439

Mohler G, Bertozzi AL, Carter J, Short MB, Sledge D, Tita GE, Uchida CD, Brantingham PJ (2020) Impact of social distancing during covid-19 pandemic on crime in Los Angeles and Indianapolis. J Criminal Justice 68:101692

Payne JL, Morgan A, Piquero AR (2021) Exploring regional variability in the short-term impact of covid-19 on property crime in Queensland, Australia. Crime Sci 10(1):1–20

Piquero AR, Riddell JR, Bishopp SA, Narvey C, Reid JA, Piquero NL (2020) Staying home, staying safe? A short-term analysis of covid-19 on Dallas domestic violence. Am. J Criminal Justice 45(4):601–635

Ratcliffe JH, McCullagh MJ (1998) Identifying repeat victimization with GIS. B J Criminol 38(4):651–662

Ratcliffe JH, Rengert GF (2008) Near-repeat patterns in Philadelphia shootings. Secur J 21(1):58–76

Secretaría de Salud del Gobierno de los Estados Unidos Mexicanos. Acuerdo por el que se establecen las medidas preventivas que se deberán implementar para la mitigación y control de los riesgos para la salud que implica la enfermedad por el virus sars-cov2 (covid-19)., marzo 2019

Steenbeek W (2018) NearRepeat: RPackage. R package version 0.1.1

Stickle Ben, Felson Marcus (2020) Crime rates in a pandemic: the largest criminological experiment in history. Am J Criminal Justice 45(4):525–536

Sun Y, Huang Y, Yuan K, Chan TO, Wang Y (2021) Spatial patterns of covid-19 incidence in relation to crime rate across London. ISPRS Int J Geo-Inf 10(2):53

Syamsuddin Rahman (2021) The effect of the covid-19 pandemic on the crime of theft. Int J Criminol Sociol 10:305–312

Vilalta Perdomo CJ (2011) El robo de vehículos en la Ciudad de México: patrones espaciales y series de tiempo. *Gestión y política pública* 20(1):97–139

Wang Zengli, Liu Xuejun (2017) Analysis of burglary hot spots and near-repeat victimization in a large Chinese city. ISPRS Int J Geo-Inf 6(5):148

Yang M, Chen Z, Zhou M, Liang X, Bai Z (2021) The impact of covid-19 on crime: a spatial temporal analysis in Chicago. ISPRS Int J Geo-Inf 10(3):152

Printed in the United States
by Baker & Taylor Publisher Services